theanatomy ofamiracle

Dr. James B. Richards

MileStones
International Publishers

Dedication

This book is dedicated to two long-time friends and personal health care providers. Dr. James G. McMurray was so patient, generous and kind through my years of struggle with a kidney disease. He kept me alive while I was learning how to walk by faith. He patiently allowed me to incorporate my faith into the healing process. He always allowed me to have room to listen to my heart and follow the voice of God.

Dr. Garry Cook has worked faithfully to keep my physical body healthy and strong. His unique insights into the working of the human body have been a constant source of inspiration and understanding for me. His insight and his dedication to physical health is surpassed only by his personal openness to the miraculous. Through every health crisis, he has worked to interweave the natural and the supernatural. He has never failed to help me keep one eye on the natural and the other on the supernatural.

If there are rewards for our accomplishments, I surely will share them all with these two men, without whom I may have never lived to fulfill my life's destiny.

Table of Contents

Introduction 9

Chapter 1. 13
    Rediscovering Childhood

Chapter 2. 19
    Grasping the Miraculous

Chapter 3. 25
    Defining the Miraculous

Chapter 4. 33
    The Logic of Miracles

Chapter 5. 39
    The Unseen World

Chapter 6. 47
    The Law of Faith

Chapter 7. 55
    Crossing the Void

Chapter 8. 63
    Shaping Our Possibilities

Chapter 9. 71
    Shifting Our Awareness

Chapter 10. 79
    Quantum Power

Chapter 11. 85
    Sustaining Your Belief

Chapter 12. 93
    Programming Your World

Chapter 13. 99
    A Miracle in Your Inner World

Chapter 14. 107
    The Miraculous Power of Being Present

Chapter 15. 115
    A Heart for the Miraculous

Chapter 16. 121
    Entering the State

Chapter 17. 127
    Releasing

Chapter 18. 137
    Shaping Your Brain for the Miraculous

Chapter 19. 143
    The Law of Harmony

Chapter 20. 149
    Harmonizing Your Outer World

Chapter 21. 155
    The Language of the Heart

Chapter 22. 163
    Heart Physics

Chapter 23. 171
    Persuading Your Heart

Chapter 24. 179
    Miraculous Mindsets

Chapter 25. 189
    If You Can

Chapter 26. 197
    It's Time to Act

About the Author. 203

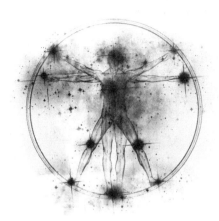

Introduction

Every culture in the world has ancient stories of miracles. These stories are widely dismissed by historians as local myths. Yet, when similar "myths" have been handed down for centuries in a multitude of geographic locations, they cannot be dismissed easily as myth. People from different parts of the globe, with no means of communication, do not make up identical stories.

In the arctic regions of North America, the elders tell stories of miraculous healings that occur to this day! One such example is a story that comes from the Arctic of a man who should have died when injured on a hunt. The snowmobiles of earlier years were not the sleek machines of today. They didn't have all the "bells and whistles"; in fact, many of them didn't even have lights. They were very heavy and quite awkward to maneuver. After launching unawares over a snow drift, this hunter's snowmobile landed on him, severely breaking his leg!

As he lay helplessly in the freezing snow, his death seemed imminent. In just a few hours the sub-zero temperatures would penetrate his arctic clothing and he would freeze. But this story, like many that are told, had a much different ending! Instead, this man employed the beliefs and practices taught by his elders. As he lay in the snow with a severely broken leg, he imagined his leg being healed as he called on the Creator. He even felt it move as all he imagined became reality. In a short time he was up on his feet resuming his hunt!

Miracles Are Real

Every culture has records of miracles. Some are of miraculous healings like the story of the Inuit hunter. Some appear to defy the laws of physics, like Qi Qong (Chi-Kung) masters of China who move objects with energy. Or like Buddhist monks who sit with nothing but a loin cloth in freezing snow covered with blankets dipped in an icy river, yet who generate enough body heat to dry the blankets and maintain their warmth. The New Testament of the Bible and other religious groups tell stories of people being physically transported from one location to another. Then there are the rain dances of the indigenous natives of North America that were said to have brought rain in times of drought. The list is endless! The evidence is undeniable!

Both sacred and secular history abounds with accounts of miracles. With the emergence of quantum physics, the scientific community is making incredible advancements in understanding the miraculous. What once was accepted by faith, intuition or empirical observation now is explained in twenty-first century language. The ancient texts and the scientific journals are sounding increasingly similar.

The ancients used metaphors of nature to explain the process of a miracle. The ancient parable employed by numerous cultural and religious groups were the only available examples that made the miraculous understandable to the mind of that day. Ironically, it is that very language, which enabled people to experience miracles in the past, unable to experience them today. Ancient language makes the idea of miracles seem archaic and "unscientific." It doesn't cause us to conceptualize as much as it causes us to dismiss.

Herein lies the ultimate secret of accessing miracles. Believing for a miracle is not about intellectual data; it is about conceptualization and integration. Therefore, the language used to facilitate conceptualization must be adapted to benefit the listener. A parable is only valuable when the language being employed has the desired effect on the listener!

The Secret to Getting a Miracle

The New Testament often uses the Greek word *logos* when referring to the Word of God. *Logos* is not so much about the literal words spoken. It's more about the logic, wisdom and integrity of the word spoken or written. While various religious groups agonize over the literal interpretation of the words in their ancient texts, they miss the point. It is not the words themselves that are so valuable. What matters is those words' capacity to help us conceptualize and integrate those words into our lives in a functional manner. The words themselves are not holy or valuable beyond their ability to move us past intellectual data and into experiential knowledge.

People must see, perceive or conceptualize a miracle before they can believe for it. The purpose of parables was to make it possible for the people of the day to conceptualize something they had no language to explain. Explaining the miraculous in a way the mind could conceptualize also enabled people to believe and conceive things as reality. This entire process was part of persuading the heart to believe for a miracle. The laws that govern miracles have never changed. It is the terminology needed to make conception possible that has changed.

Quantum science is using a new language to describe the same phenomenon. The metaphoric language of the twenty-first century must be a terminology that simplifies the inner working of the miraculous just as the language of farming, shepherding and current events did for the mind of earlier days. Parables that talk about growing wheat are of little value to the average person in the modern world, especially when he[1] has never seen a wheat field or worked a garden!

The world, while teetering on the verge of destruction, is simultaneously entering into a new day of miracles. Cutting-edge, scientific discovery is giving twentieth century terminology and scientific explanation to what the world once knew only by intuition or empirical observation. We are living in a day when science and miracles are finding common ground, narrowing the once enormous gap to little more than a semantics issue.

The Anatomy of a Miracle puts that which has been known for thousands of years into a language that is readily understandable to the twenty-first century mind, giving us the opportunity to develop a faith, a deep knowing, that we can move into the invisible realm and come back with a miracle.

---

1   For the sake of simplicity, I will use the old-fashioned rule of "he," "him" and "himself" to refer to both male and female humankind.

Chapter 1

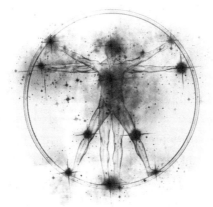

Rediscovering Childhood

Deep in the heart of every human being, there was, at one time, an innate longing to see and experience the supernatural. Unfortunately, like all children we were taught to "grow up." Growing up should be the development of our inherent God-like qualities that seem so natural and logical to children. Instead, society, in nearly every culture, has made growing up a denial of who we are in relation to our Creator, how we were created and how we should function. For most of the world, adulthood is the denial of all that is unique and inherent to us as the children of our Creator!

Those of us adults who do continue to embrace some notion of our divine identity, instead of making it simple and natural, make it difficult and out of reach. We have somehow come to believe that those gifts and inexplicable abilities that belong to all men are relegated to the few who will adhere to some religious discipline, ascetic lifestyle or code of perfectionist conduct. And true to our nature, if we believe it is hard, it is! If we believe it is beyond our reach, it is! If we disqualify ourselves, we are disqualified! To quote Henry Ford: "Whether we think we can or whether we think we can't, we are right."

The greatest miracle worker of all time never taught how hard or difficult it was to experience the miraculous. Instead, He expressed contempt for those who made God's provision difficult and out of reach. Jesus took the position that making something harder to get was tantamount to taking it away from people. Instead of trying to make people live up to unrealistic codes or participate in religious or ascetic practices, He simple urged them to, "Only believe." The written record of His miracles, which was only a small portion of what actually occurred, happened with no great fanfare or religious observations.

Possibly in our process of becoming adults we have lost the divine sense that fueled our imagination as children. Maybe our intellectual mind has so dominated our every perception that all things that cannot be intellectually explained are simply washed from our consciousness. Perhaps we have believed a lie and made everything about life too difficult. Or maybe, as the greatest miracle worker of all time said, we forgot how to think like children.

We have forgotten how to think like children.

If connecting to the miraculous requires our rediscovering of who we came into this world intended to be, we each have to ask ourselves, "Will I allow such a thing? Can I give up so much?" We have spent our entire lives developing a perception of the world that makes us seem intellectually superior. Medical science ignores cutting edge, concrete data, instead preferring outdated, disproved models that serve only to line the pockets of the drug industry. We have defined rules of science, yet many of those rules only exist in non-provable theory. We have fought to suppress information that exposes the possibility that we can be more than that which serves a global economy. The accepted worldview is one that facilitates our inflated egos, which has grown disproportionately enormous as we have passively surrendered our true identities. Our entire lives, from the time of becoming adults, has been dedicated to an illusion of reality.

Children have a much greater need to enjoy life than to protect ego and win arguments. Children don't need intellectual explanations for what works. They don't care why the light comes on when they flip the switch. They just don't want to be in the dark. It is of little concern who gets the credit for building the toy; they just want to play with it. Children have little concern for who owns the toy as long as they get to play with it. In order to become like children we must be willing to accept what is, whether or not it fits into our life-view.

The adult in us wants to compare our current beliefs and opinions to those embraced by others. Before considering what works or what will bring quality to our lives, we want to know if accepting another's view will make us wrong. Right and wrong are based on intellectual rigidity with no allowance for relativity.

For thousands of years mankind lived and operated, evidently, fairly successfully on the oral traditions held by the most ancient of civilizations. There was little need for scientific explanation. It just needed to work! No doubt there were many superstitions and religious ideals that didn't work. Unfortunately, they usually were accepted based on the persuasiveness of the "soothsayer" who benefited from the idea.

In time religious dogma and tribal superstition exchanged that which worked for that which gave control to the few. What was consistent with creation, or what the Chinese called the Tao and the Greeks tagged the logos, was ignored. The wisdom of the Creator was exchanged for the logic of ego-centered, power-hungry men. This departure from what worked drove man into another extreme: science!

Modern science, to some extent, was a reaction to the superstitious mindset that ushered civilization into the Dark Ages. Like all reactionary thinking, it threw out too much. It rejected the good and the bad. In an attempt to bring logic and order, a new extreme was imposed upon man that led him away from his natural God-given rights and benefits. This new science sought to explain the world in a framework that simply put a new group of power brokers in control.

They didn't wear headdresses, call on spirits or chant. They adorned themselves with suits and were affluent and educated. Yet, try as they did, their educations were as limited as the ancient shamans they sought to eradicate. The former tried to understand the world by subjectively interpreting what they *couldn't* see. This new group tried to convince themselves that the only thing that existed was what they *could* see!

Newtonian physics became the crown jewel of logic and explanation of all things. Because there were so many things right about these mathematical equations, it was assumed that nothing could be added or taken away. This was the "be all" and "know all" end to all questions. All the questions of the universe were naively interpreted in a manner that would preserve and protect the supremacy of the accepted logic.

Man, as an adult, has an either-or attitude. Ego and agendas are so closely connected to ideas that there is little room for growth, understanding or even perception beyond what supports our accepted views. History has proven that much of mankind would rather die clinging to an untruth than embrace a new truth and live. So laws, religions and governments are established to protect what we have accepted, whether it works or not. Now that we all agree on a common idea or belief…we are adults!

No New Truth

Accepting a new reality has nothing to do with right or wrong. Why? It is because there is *no new truth*. All truth always has existed. It was here before we were born. It will be here when we depart. It works for others regardless of our opinions. It exists where we see it and where we don't. It isn't new. It is only new to us! Yet, there is something in us, as *adults*, that makes us feel inferior when we identify a truth that we have not yet understood. We feel that to say, "I didn't know," is to say "I am flawed." We think that to expand what we currently believe is to refute what we previously believed. We are relegated to a life of limitations because of the fear of being wrong.

Truth isn't new. It is only new to us!

I've got great news for you. To move into the realm of the miraculous has little to do with being right or wrong. For example, as I stood at the base of a great mountain range in western America, it looked as though the mountains touched the sky. From the base of the mountains it looked as though there was nothing that existed beyond what I could see. As I drove up the mountain and reached the peak, I saw for the first time the incredible view that had just hours before lay beyond my sight. What I saw from the mountain's peak did not deny what I saw from the basin below. It was simply a new view from a new vantage point.

Newtonian physics is evidently very accurate when dealing with the "seen world." It gives us the equations to launch space vehicles, design aircraft and pretty much conquer the "seen world." But there is an *unseen* world, a world that lies beyond the explanation of Newtonian physics. That world is governed by an entirely different set of rules and equations.

The subatomic world has come to be understood through a science called quantum physics. Quantum physics is a science. It is a science, however, that touches on the unseen—that part of our world that previously had been consigned to the religious and the superstitious. Quantum physics is able to help us identify the truth of some of the oldest religious teachings in the world. It is able to give us a language that makes sense to our generation and explains what happens in the invisible world when we apply spiritual truth.

It is my desire that this book will open your eyes to such an extent that you will open your heart to a life of possibility, available to all, given by our Creator. In so doing, you will enjoy the very best life has to offer. What others call miraculous will be your normal. There will be no need of a declaration of being wrong. There will be no new truth. There only will be the willingness to look at the world from a new vantage point that opens an entirely new world of limitless possibility. Like little children, we will have little concern for who was right and who was wrong. We simply will enjoy the goodness of God in a world that is designed for the miraculous!

Chapter 2

Grasping the Miraculous

The very word *miracle* tends to conjure images of mystical apparitions and unsearchable knowledge that reaches beyond the comprehension of the human mind! For many it is a "reaching beyond the veil" into a mystical realm of the unknown. Ironically, there is a seemingly equivalent vain attempt by others to define the miraculous in ways that make it fit into finite bounds of intellectualism. Regardless of which extreme facilitates our particular comfort zone, both make it virtually impossible to *experience* the miraculous.

An ancient proverb says, "Open rebuke is better than secret love."[1] The point of the saying is not to emphasize the value of open rebuke. No one really values or cherishes an open rebuke. It is humiliating and demeaning at best. However, secret love, or love that is not expressed and experienced, is of less value than the humiliation of an open rebuke. Likewise, a miracle not experienced is worthless. Formulas and rituals do not meet the real need. They simply serve to pacify the mind that has not yet experienced the reality!

---

[1] Proverbs 27:5 KJV.

Another ancient proverb says, "The way that can be explained is not the way." The Chinese recognized a way in which life and energy flowed so easily that it allowed things to occur as they should. They called it the Tao. The Tao, or the way things naturally worked, could be grasped by observation and contemplation. In their wisdom they understood that any attempt to over-intellectualize "the way of the unseen" made it impossible for the heart to conceptualize it. After all, the goal is not to explain the miraculous but to experience the miraculous!

Extreme mysticism or intellectualism is laced with excuses for failure. The mystical approach to the miraculous has no end to the rituals, religious duties or secret revelations required to experience the ultimate. But in the end, the overly mystical person seldom experiences the miracle he pursues! He always needs one more qualifying factor!

Likewise, intellectuals substitute formulas for experiences. When the need to be right supersedes the need to be whole, intellectuals will produce a great formula that can be outlined, taught and intellectually explained. However, information is not even a close second when someone needs a physical healing or emotional relief! Intellectualism, like mysticism, has an endless assortment of built-in reasons that occupy our minds while we languish in the perpetual state of lack.

Our Hearts Know

Mankind seems to live in an incredibly conflicted state. At a heart level we know that miracles should be a way of life. Yet, intellectually we are so sure we will not experience the miraculous that we inject our excuses for failure into our system of approach. We have an excuse or explanation ready for every disappointment. It seems that having one more step to take or one more piece of the puzzle to find keeps us from facing the devastating sense of emptiness that would swallow our hearts when we are not able to experience the miraculous.

Sadly, we have never reconciled with our inherent nature. The realm of the miraculous has been condescendingly relegated to the mystics and the religious. At the same time, no man has escaped his inner nature. We were created in the likeness and image of our Creator. On some level we all pursue the miraculous. The scientist quenches the yearning by seeking to discover the next scientific breakthrough that will liberate man from the bounds of time and space. The medical practitioner seeks the cure of a deadly disease through allopathic medicine. The philosopher attempts to improve life by freeing the mind. All of these are but intellectually stimulating, socially acceptable, egotistically palatable attempts to satisfy the inner certainty that we can live in the miraculous!

Man is internally governed by *beliefs*. Whether our beliefs are right or wrong, they still govern our lives and our capacity to experience life. As Henry Ford so aptly said, "Whether you think you can or whether you think you can't, you're right!" Beliefs make anything possible or impossible to the person who holds those beliefs.

> ## *"Whether you think you can or whether you think you can't, you're right!"*

Jesus, the Great Healer, said it like this: "All things are possible to him who believes."[2] "All things" does mean ALL THINGS! Evidently the greatest healer of all time clearly meant that this law applies to every situation. Thus, it should go without saying, that those who don't believe find impossible to them things that are clearly possible to others. There is no denying the miraculous. Every day doctors see people recover who should

---

2   Mark 9:23 NKJV™.

never live. When they see a miracle they often call it the placebo effect or some other equally dismissive term. Scientists make new discoveries that defy all we have understood about the world to this point. These very occurrences that are so conveniently explained within the context of their finite understanding are things that happen because someone believed.

Now Our Minds Can Know

Cutting-edge science has finally advanced to a state that puts the realm of the miraculous within the grasp of modern man. The language used by healers and miracle workers of the past to convey deep truths was relevant to the logic of their day. Today's pseudo-intellectual seizes upon the seemingly scholarly limitations of the metaphors used in a past era and discounts the truth. However, closely examining the scientific discoveries of the past decades in the field of quantum physics reveal what the Great Healer meant when He spoke in parables.

The concept of living by faith is no longer a leap into the unknown.

The concept of living by faith is no longer a leap into the unknown. Faith is a rational step into a somewhat known realm, one that has been experienced by the spiritually minded for millennia and now has been scientifically verified. When Jesus talked of living by faith, or when He made statements like, "All things are possible to him that believes," He was cutting to the heart of what happens when a human being operates by the immutable laws of creation.

The cloud of mysticism that was created by extreme religious interpretation is removed when the words of the Bible are expressed in modern terminology. Scientific terms that are more easily understood in the twenty-first century mind help us to understand the words spoken by mystics, healers, prophets and Jesus Himself!

Milton Erikson was famous for his ability to use hypnotic language. He would tell a story filled with metaphorical concepts whereby people would grasp a truth and subconsciously apply the principles to their personal issues. Even though the client may not realize what was happening, as he lost himself in the story, his heart would grasp a truth that caused a miraculous change.

This is exactly what Jesus did with parables. He used the language of the day. He told stories about things people understood, such as farming, shepherding, vineyards and the weather. He used allegory to get people to subconsciously draw parallels between divine principles and natural occurrences. Having removed them from the realm of the mystical and the intellectual, they were able to enter the realm of believing and thereby experience the miraculous.

When those people experienced the miraculous, they could tell what happened, but they could not explain how or why. Like anyone in a desperate situation, when we experience the miracle we need, the how and why is of little concern! The *experience* is everything!

Believing is not an intellectual happening. It is something that occurs at the heart level. It is something that happens so deeply that it affects the way we think and feel. It alters the way we see ourselves. It happens instantaneously and without effort. It can be grasped but not learned. It can be experienced but not earned!

Don't over-intellectualize this book. Another ancient proverb says that if you think you can find the answer in a book, don't read the book. If you think a teacher can give you the answer, don't get a teacher. Books and teachers only encourage and spur you on. In the end it will be what happens in your heart that connects you with your miracle!

Chapter 3

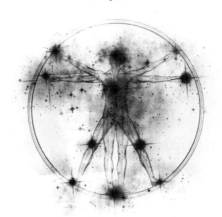

Defining the Miraculous

What we call the *miraculous* can and should become a way of life. Ironically, something as simple as our definition of the miraculous may be the greatest deterrent. Why? It is because definitions create interpretation barriers. Our interpretations are the filters through which our minds process all related data. They determine the way we understand any information. Therefore, an inaccurate definition can alter our perception and pursuit of any aspect of life. The simplest thing can become impossible in our minds by nothing more than an incorrect definition.

Many of our deep beliefs were never developed by conscious deliberation. They were imposed upon us by our culture: our family culture, our religious culture or our regional culture. Cultural beliefs are the most difficult to resolve. Since they were never accepted by intelligent choice, they seem to be an inherent universal truth accepted by all. To challenge a cultural belief puts us in opposition to all that seems natural to us and the world as we have come to know it. Regardless of where our definition was conceived, it often can be a subtle concept that limits our capacity to participate in the miraculous.

To understand miracles we must endeavor to look at them internally; thus the title, *The Anatomy of a Miracle*! Anatomy has to do with the structure, framework or internal parts. By looking as much as possible at the inner workings of the miraculous we can remove the mysticism that clouds the reality. We must get beyond subjective explanation and emotional testimonials. Although these can have great value for encouragement, they can never be our source of understanding the inner workings of the miraculous!

When people experience the miraculous, they can explain it only within the realm of their perceived experience and the clarity of a shared vocabulary. A shared vocabulary is when two people share the same concepts and definitions of words and phrases. Seldom do two people mean exactly the same thing even when speaking the exact same words. This is why the miraculous cannot be fully explained.

Explaining a miracle is like painting a beautiful landscape. It is at best a subjective representation of the reality. The person painting the scene can interject all manner of personal preferences and interpretation into the painting. And if another painter actually experiences the same scene, his experience will be different, thereby creating a slightly different painting. Or as it is said in neuro-linguistics: The map is not the territory! Many times what is subjectively perceived is not based in reality. It is simply the only way of describing the event within the scope of our vocabulary and life experience. It is often these subjective explanations that drive us away from the validity of the miraculous.

Explaining the miraculous could be likened to an uncivilized cave-dweller attempting to explain how someone was cured of a life-threatening illness with a shot of penicillin. If such a person had never seen a doctor, a doctor's bag, a needle, gauze or any of the other associated paraphernalia, he would have no shared vocabulary to explain what happened. His language would not have words to explain the subsequent biological functions. He would use such antiquated, inadequate terminology that even when explaining what happened to a medical doctor, it would seem like nonsensical superstition. Especially if the doctor didn't realize the cave-dweller was describing a simple shot of penicillin!

Inherent Power

Looking at the ancient Greek language used in the New Testament, the word *miracle* was defined as "inherent power, power residing in a thing by virtue of its nature, or which a person or thing exerts and puts forth."[1] Accepting this definition of a miracle might force us to reexamine and redefine our accepted definitions and concepts. The concept of inherent power is foreign to most, but it is essential for a life of the miraculous.

For too long the general definition of the miraculous has been something like this: "an unexplainable occurrence that defied the laws of nature." Subjectively, this is what we see and experience. But the internal working of the miraculous, according to the previous definition, may be something altogether different.

"Why all this fuss about a definition?" you may ask. What you believe not only drives and shapes your experience, it also is the first key to unlocking your capacity to perceive and believe! This definition, as embraced by Jesus the Great Healer and supported by cutting-edge science, opens the door to new possibilities of living and moving in the miraculous. It eliminates the mysticism and religious dogma that alienate people from what may be our inherent right as citizens of planet Earth and children of the Creator!

---

[1] The New Testament Greek Lexicon <<http://www.studylight.org/lex/grk>>, *dunamis*. November 22, 2008.

When you understand the inner workings of any process, all you have to do is duplicate that process to get the same results. By understanding the inner workings—that is, the anatomy—of a miracle, we open the door to a life where all things really are possible, all of the time!

Inherent power is an essential concept in grasping the miraculous. The world and all that is in it was brought into being and exists by inherent power. Quantum physics tells us the universe cannot exist apart from the interaction of intelligent life. Therefore, it stands to reason that it was brought into existence by intelligent life. Regardless of what you believe about the creation of the world, the laws that govern all things still apply. You must choose how you will interpret and apply the governing laws, but you cannot ignore them and still live in the miraculous.

> *Regardless of what you believe about the creation of the world, the laws that govern all things still apply.*

There is a constant interaction between all intelligent life and all that exists. It does not require faith for this to happen. It just happens. It is the way of all living things! Although faith, or believing, isn't required to bring about the interaction, faith may be essential to believing that we can make choices concerning the influence and outcome of that interaction.

So if we are able to make deliberate choices about how we will influence the world within or around us, then it is conceivable that we could bring about a miracle. This would, of course, require that we know and harmonize with the laws of the miraculous. But before we go much further in this idea of interacting with the miraculous, let's consider some inaccurate definitions.

Debunking Our Definitions

The first inaccurate definition of a miracle says, "A miracle is what occurs beyond the scope of explanation." This may be one of the most universally limiting misconceptions. There are things that happened hundreds of years ago that were categorized as miracles solely on this definition. However, when scientific discovery evolved to understand and explain the process, it was "declassified." If a miracle is based on our ability to explain, then a miracle is subjectively defined by the intelligence of the participants. It is this definition, at least in part, that has closed the scientific community to the existence of miracles.

The second inaccurate definition says, "A miracle is something that violates the laws of nature." Based on this concept, an airplane in flight is a miracle; that is, until you come to understand higher laws of nature (physics). There were times when flight would have violated the known laws of physics. As other laws of physics were learned, man could, through a complex combination of laws, do what previously seemed impossible.

Through the field of quantum physics we are beginning to understand laws of nature that have heretofore been completely hidden. In fact, quantum physics has become the greatest supporter of ancient truth espoused by the Bible and many other ancient texts. For the first time, that which occurs in the realm of the invisible has become understandable. Now that it can be explained in a twenty-first century vocabulary, it becomes believable to modern man. For the first time in history we see that science and religious belief does not have to be antagonistic. Many of the seeming differences are simply a matter of vocabulary!

The third inaccurate definition is this: "A miracle is what occurs when God takes sovereign action." This is by far the most destructive concept. Such a definition leads to conflicting concepts of favoritism or even religious legalism. Whether looking at the Old Testament or the New Testament or at some other concept of God, the idea that man has freedom of choice and, in fact, has authority and responsibility in planet Earth is clear and undeniable. This wrong definition violates every biblically accepted concept of God, a topic that we will explore in future chapters!

Quantum physics has become the greatest supporter of ancient truth.

Last but not least is this false concept: "Miracles are the exception." The moment something becomes the exception, it eludes the reach of the masses. For anything to be the exception makes it unpredictable, indefinable and unattainable. If we believe there is an exception to a law of life, we will always believe it applies to our situation. That concept makes our situation irresolvable, or at best incredibly difficult.

So, what would be a reasonable definition of the miraculous? What definition would be accurate and still facilitate our capacity to grasp and experience a miracle? Although this is by no means meant to be the absolute definitive definition of a miracle, let's use this for now: "A miracle is when we harmonize with a law of nature that releases the inherent power of life that resides in all things."

Using this as a definition, we free ourselves from the burden of feeling that we have to generate the power by our exceptional faith. It frees us from the notion that we must somehow convince God to act in our behalf. It frees us to make the choice to harmonize with the laws of the miraculous and experience what is freely given to all men!

Chapter 4

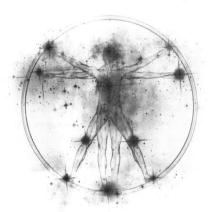

The Logic of Miracles

The most challenging factor confronting our entering and abiding in the realm of the miraculous is its utter simplicity. For an overabundance of reasons, people have put miracles in a realm that made them far away and unattainable. Accepting their simplicity and easy availability will challenge a lifetime of accepted, unquestioned but limiting beliefs. It will cause us to face and fight our greatest obstacle: the need to be right!

If a miracle happens when I harmonize with a law of nature, then the possibility presents itself that once I learn these laws and recognize the process, the miraculous can become a way of life. That is exactly what I am presenting to you! You are entering into a realm that was once thought to be for a few elite; a realm for those of such great faith or those of such moral perfection that it seemed unattainable to the average person! Now the realm of the miraculous can be a way of life for you!

The Chinese called it the Tao. The Greeks called it the logos. The Bible calls it the wisdom of God. Nearly every culture and religion has a term for it. What is the "it" to which I am referring? Modern science calls it physics. However, regardless of the terminology, regardless of the inaccurate associated beliefs and theories, nearly all groups and cultures recognize this fact: The world and all that exists operate by a set of consistent, predictable laws.

One almost never would put physics and miracles in the same conversation. The two seem exclusive of each other! Yet it is the realm of physics that has finally unraveled much of what we can know about miracles. For the first time in the history of the world, man has been able to see the "unseen world" and discover many of the laws that cause that world to function. What was cloaked in antiquated, often religious language now can be understood in a language and context that makes sense to the modern mind.

Physics: The New Vocabulary

Two thousand years ago a New Testament writer said that the things that are seen are made from the invisible,[1] a concept that until now could be grasped only in some esoteric ideology. Now a science exists that explains what was once beyond logical explanation. We now have a shared vocabulary that makes it possible for us to discuss the invisible world in an understandable way. Today I can talk about the invisible world in a way that makes sense to the Christian, Jew, Buddhist, Taoist or scientist without reducing it to a theological debate that stalls in the quagmire of nebulous dogma. Regardless of the individual terminology employed, with this shared vocabulary we're all talking about the same thing and we all should be able to agree that we are talking about the laws of nature.

The miraculous does not occur when we violate the laws of nature. The miraculous occurs when we *harmonize* with those laws. Even when the law cannot be explained or is unknown, it is still a law of nature. The

---

1   See Hebrews 11:3.

very laws that brought about air travel, television broadcast, heating and cooling and every other modern convenience are laws that can bring about what once was considered impossible! But if history has proven anything, it is this: What is impossible to one generation is common occurrence in another.

The miraculous occurs when we harmonize with the laws of nature.

What the Greeks called the logos existed with the Creator. Although the word *logos* can mean many things to many people, it embodies the idea of a logic or reasoning. The writers of the New Testament implied that all that was created was done so based on a particular logic. That logic was upheld by the character, consistency and wisdom of a loving Creator. Jesus, the Great Miracle Worker, demonstrated what could happen when we follow His example and harmonize with the inherent laws of nature. Whether it was Jesus or others who worked miracles, they all worked with the same laws of physics.

The wildly successful book *The Secret* is a perfect example of someone applying an idea that can be acceptable to Christians, Jews, Buddhists or scientists if explained in understandable terminology. They called it "the law of attraction." Regardless of what we call it, it is based on a law of nature that has a plausible scientific explanation. Hundreds of people from all faiths have worked miracles, not because of their particular religious beliefs, but because of their ability to identify and harmonize with one of these natural laws. Millions of people, with vastly different beliefs and backgrounds, have successfully applied "the law of attraction" within the context of their beliefs. These laws are inherent and universal.

The insistence that we all use the exact same terminology and embrace the exact same religious or scientific ideology has done nothing but push the world of limitless possibility even further from our grasp. It is this separation that has alienated man from the miraculous. Regardless of how we explain the beginning of the world, it exists by a set of definable and somewhat predictable laws. Making those laws work for us does not prove or disprove our ultimate concepts of God, creation or science; it simply verifies that the laws always work regardless of other belief factors.

In fact, it is a Christian idea that says we can understand the invisible attributes of God by all the visible things He created.[2] This idea moves God from the realm of the unknown and unpredictable. In fact, this would mean that a miracle can never defy a law of nature; if it did, it would make God impossible to know and predict. One of the characteristics separating the great miracle workers in the Bible from the common man was often the fact that they had moved beyond simply observing God and had entered the realm of knowing God. We can know so very much about Him by what we understand about nature…physics.

Know-It-All Thinking

As a young boy growing up in the 1950s, I was faced with a world that seemed to be ever changing. Doctors and scientists made definitive statements about the world and the human body, only to discover that their ideas were wrong or limited. They couldn't, or wouldn't, write new textbooks fast enough to keep up with the newly discovered changes. It wasn't, however, the world that was changing; it was our understanding of the world. This information opened doors and forced upon us new paradigms that have forever changed people's understanding of the world.

The same scientific community that was so sure it had all the answers in the 1950s is ever so predictable. It seems no matter where we are in our

---

2   See Romans 1:20.

current understanding or how many of our previously held theories have been disproved, there is an arrogant insistence that our current view of the world and human existence is the consummate revelation. Our search for security in our ideas and opinions closes our eyes to a world of miracles that are ours for the taking. But we live in a day when scientists, philosophers and religious leaders are opening their eyes to what is the right of every human on planet Earth: miracles!

With all that we see and know, we must ever be aware there is so much more. There is within every human being the desire—no, the need—to experience the miraculous. Regardless of what we know or believe at this point, there is always more—and it is always better than we can imagine. We must stay open to new insight, revelation and scientific discovery. We must accept nothing that makes us feel limited. We must enter the realm of possibility by committing to an attitude that says, "There are no limitations; all things are possible!"

Within every human being is the need to experience the miraculous.

The religious thinkers fear that the idea of an established set of laws will diminish man's desire to know his Creator. The truth is, it is the idea that God is good and yet we have to run to Him and convince Him to perform miracles that is a contradiction. These conflicting beliefs have turned away multitudes. It has engendered the unanswerable question, "If God is so good, why do some get miracles and some don't?" We have this erroneous religious idea that the Creator must make an independent decision for every need presented by man.

Instead it seems that God created a world that worked by laws that were accessible and understandable to mankind. He has made those laws known to every generation as much as they were willing to consider knowledge beyond their opinions and dogma. It seems that He is so loving and so benevolent that He gave us a capacity held by no other species. We have the capacity to choose and change our quality of life by making decisions and applying the universal laws of life.

Those who embrace and experience those laws have an incredible appreciation for such a loving Creator, who has richly and freely given us things for our enjoyment.[3]

---

[3] See 1 Timothy 6:17.

Chapter 5

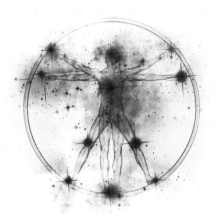

The Unseen World

Entering the realm of the miraculous is entering the realm of the unseen. In modern language the unseen world is the subatomic. This subatomic world, although more concrete than the seen world, exists beyond the grasp of the five senses. Yet, it is not beyond the realm of knowing and experiencing! It is visceral; it is known beyond the awareness of the tangible. Therefore those who rely totally on the five senses enter this realm—if they enter at all—with disbelief and trepidation! This disbelief is not about this realm's existence; it is disbelief in the laws that govern this realm.

Our relationship to the world, as it is today, is completely inside out. Because of our over-reliance on the five senses, we have become carnal. We seek to know and relate to this world within the limiting scope of our physical awareness. We erroneously hope that mastering the world that is seen is mastering the world, period. Nothing is further from the truth. To master the seen world is tantamount to thinking that sailing a ship is mastering the sea. We have, at best, learned simply to navigate what we know of the world.

In the beginning man relied on his innate knowledge of the unseen world. The Hebrew Bible says that the Creator gave man authority over the world.[1] That authority was not "rule by force." It was ruling the seen by the unseen. It was an innate awareness of how the universal laws worked. Having sprung from the Creator, man's internal knowledge was his source of power and authority. It was only as man departed from his sense of true self that he substituted power with force.

Man's Journey From Inner to Outer

From the first man, the knowledge of the irrefutable laws of nature was passed along for generations. Consequently, all cultures have a fundamental body of knowledge and beliefs around a similar premise. As man moved away from the unseen and grew more reliant on his natural senses, he lost touch with what was rightfully his as a child of the Creator: his ability to exercise authority and rule the seen from the unseen.

With the need to rule as part of the fundamental makeup of man's nature, his perversion of life substituted power with force. As violence became the mode of exercising force, man surrendered his trust for the unseen. His need for survival in a world that had become violent forced him to rely on the natural world. Throughout subsequent generations the masses lost the awareness of the invisible. Fortunately, throughout history a few sages and holy men preserved as hidden truth what once had been known to all men.

As the masses lost the ability to live in the inner world, a vacuum was created. As man became less intuitive, his reliance upon the intellect abounded. Inventions became the external replacements for internal realities. These inventions were not wrong; they were essential for survival. In the absence of man's true capabilities, those needs had to be met in some form. Although they were a testament to man's intellect,

---

[1] See Genesis 1:26,28.

these inventions were a poor substitute for what could have been. These inventions merely were vague images of what man intuitively knew to be his divine capacity.

With the emergence of such conveniences, the need to cultivate the inner man became less and less attractive. Expediency defined the road of least resistance. Convenience required no character or cultivation. Man, who earlier had surrendered his inner power to violence and force, now traded away his inheritance for convenience. While a few continued to stay in touch with the higher laws throughout the centuries, the masses dulled their inner senses to the availability of inherent power and universal laws.

In ancient times there was no language to describe what occurred in the unseen. It simply was understood. The Chinese, whose history can be traced back to the Garden of Eden, mastered the art of empirical observation.[2] Although they had no words to describe what happened at the subatomic level, they became very comfortable with the fact that unexplainable phenomenon occurred in a somewhat predictable manner.

The early Taoists were the first true scientists. They knew the world had been created and was sustained on certain laws. They called that the Tao (pronounced *dao*). They understood that these laws could not be manipulated. However, they also knew that the predictability of these laws could give man the capacity to harmonize with them and improve the quality of his life.

From the Chinese Taoist to the American Indian, on different continents, in different languages, within the context of their understanding, each culture excelled or diminished in their knowledge of the laws of the universe. They gave them different names, they each developed their own set of rituals and customs and in their imaginations they each associated those laws with different deities. Some turned them

---

[2] C. H. Kang and Ethel R. Nelson, *The Discovery of Genesis: How the Truths of Genesis Were Found Hidden in the Chinese Language* (St. Louis, Missouri: Concordia Publishing House, 1979).

into secret knowledge that gave them seemingly magical powers. Others shared them openly with the masses. Some used them to serve mankind; some used them to dominate mankind. Regardless of how they were used or in what context they were applied, they were, in fact, the immutable laws of the universe, which were brought into existence to form and sustain the physical world.

> *The immutable laws of the unseen world formed and sustain the physical world.*

Sadly, it was through the diversification of application that these laws became categorized as good or evil. The New Testament presents an interesting concept that even those who claim to follow Jesus' teaching seem to overlook. Nothing is inherently evil. People make things pure or evil by their own minds, purposes and intentions.[3] All cultures invoke the same universal laws put into existence by the Creator. Whatever we believe about the Creator, it does not make what has been created good or evil. How we use it makes it good or evil!

Religion, superstition and intellectualism all play an equal part in our hysterical flight from our rightful place as citizens of the planet and our irrefutable inheritance as children of the Creator. We must accept the truth that the universal laws governing the unseen world are neither good nor evil. We must choose to make them good by the way we use them. We must not become worshippers of these laws; likewise, we must not assume that proper application of these laws make us right or wrong in our other

---

3   "Everything is pure to those whose hearts are pure. But nothing is pure to those who are corrupt and unbelieving, because their minds and consciences are corrupted" (Titus 1:15 NLT).

beliefs. We simply must accept them as they are and apply them as they work with no extrapolation into other dimensions of life.

Finding the Unseen

One of the first fundamental laws of the miraculous is this: *That which is seen is made from that which is unseen.* This is a universal truth. It is embraced by every world religion; it is clearly stated in the New Testament; and it is scientifically validated.

This then is followed by another fundamental law of the miraculous: *To master the physical world we must master the unseen world.* Wallace Wattles, one of the American forerunners of positive thinking and the first great pioneers of faith, says it like this: "To think according to appearance is easy; to think truth regardless of appearance is laborious, and requires the expenditure of more power than any other work man is called upon to do."[4] The "appearance" to which he refers is the seen world; it is that which currently exists in physical form, which we are experiencing with our five senses. The "truth" to which he refers is that which exists in the unseen, which has preeminence over the seen and has the power to change the seen; it is that which we are not experiencing in the realm of our physical senses.

It seems that the Creator imposed no limitations as to how we can come to know the truth of the unseen realities. The goal is that they be used for our good. When we experience their ability to improve our quality of life, we experience the love and generosity of a Creator who made it possible for every man to live an incredible life. But the unseen world, whether understood from a religious perspective, a metaphysical concept or scientific data, presents one fundamental limitation: It can be accessed and experienced only through specific laws. It is these laws that you will come to know in this book.

---

[4] Wallace D. Wattles, *The Science of Success: The Secret to Getting What You Want* (New York: Sterling Publishing Co., Inc., 2007), 29.

Many scientists look at the subatomic world through a microscope. They have physical proof of its existence, yet they may never experience its miraculous potential. Seeing its existence through a microscope does not equal knowing it in one's own heart and experiencing it in one's life. To observe it with a microscope is merely an expansion of the five senses. Likewise, a religious person mentally may accept the existence of the unseen world. But an intellectual recognition of a miraculous power is not equivalent to experiencing that power.

Regardless of the terminology used to describe the unseen world, whether religious or scientific, whether sacred or secular, we can by some clearly defined universal laws access and influence the unseen world in such a way that it affects the seen world. This is what for thousands of years has been called a *miracle*! Today we can remove that miracle from the realm of the vague and nebulous to the realm of experience and physical phenomenon.

Miracles are your right and your inheritance!

Today you hold the key to a life of the miraculous. Regardless of your religious background, whether scientific or sacred, you are peering into the world of the unseen and grasping what has been hidden for millennia and is now made understandable in a modern language. Miracles are your right and your inheritance! Invoke the first law of interaction with the miraculous: personal decision. Decide to live in the miraculous and, as you move through these pages, your heart will lead you beyond a life of limitation to a life of the miraculous. Don't try to figure out how it will happen. Don't evaluate the possibility of the future on the certainty of the past. Decide what will be!

Desire is a possibility ready to happen; frustration is a possibility unfulfilled. Your desire for more and your frustration of what has been is the greatest proof that your heart is ready to make this journey. You need no greater motivation than the desire for better and frustration with what you have accepted up to this point. Now you must decide. When you set your intention to break free from the limits of the known and venture past the horizon of a life of average, all the powers of creation will come to assist you.

You found this book because you are ready. The Chinese have a saying, "When the student is ready the teacher appears." The Hebrew Bible says that wisdom calls to you and pursues you.[5] Today you have heard that call. Your teacher has shown up in the form of a book. I am not the teacher. I am only pointing you to the Teacher. He will guide you past the information of these pages to the reality of the truth. Miracles are yours! Enjoy your new life!

---

5   See Proverbs 1.

Chapter 6

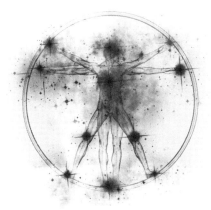

The Law of Faith

The law of faith is a universal law. Faith is one of those "common" words spoken by nearly everyone, yet for which there is no shared vocabulary. The word *faith* means many things to many people, both positive and negative. Thus it is hard for people of different disciplines to have a rational discussion about faith without continual misunderstanding.

It is difficult to have a rational discussion utilizing a word with so many contrived meanings, each of which threatens the philosophical position of another. To the intellectual, faith may be viewed as the irrational, unscientific delusion of the religious fanatic. For the religious, there are as many concepts as there are faiths and denominations. The most commonly accepted one is the idea of believing something to such an extent that God is moved to action in behalf of the believer.

Like all words that evoke such extreme emotion, the truth is never found in those who venture toward excess. In order to grasp the universal law of faith we must set aside any preconceived ideas—at least while reading this book. We may choose to hold those ideas for other arenas, but to grasp the principles set forth in the rest of this book, we must establish a shared vocabulary.

Faith Is...

For our purposes we will use the definition of faith as it was understood in the common Greek language: "to be convinced or persuaded."[1] As you can see, this word, even as it is used in the Christian New Testament, has no religious meaning. In fact, upon close examination we see that the God of the Christian faith calls its participants to believe in the unseen world.[2]

It is here the New Testament takes a leap into quantum reality when it explains, "The things which are seen were not made of things which are visible."[3] When we take away all the extremes of religious connotation, we recognize a deeper reality in this passage. All disciplines of faith have known for thousands of years what the scientific community only recently came to accept: The world we see was brought into being and is sustained by the invisible. We now have several names we call that invisible world: the subatomic world, the quantum field or the mind of God. Regardless of the terminology employed or any derived conclusions, we all are speaking of the same thing.

It is faith, the persuasion of such a field and its bearing on the physical world, that gives us the first glimpse into how we can interact with that world to bring about change upon the world we see. If a house is built and held together with wood and nails, then it stands to reason that we can change the appearance of that house by rearranging the wood and nails. At the risk of over-simplifying it, this is what happens when we act upon

---

[1] Biblesoft's New Exhaustive Strong's Numbers and Concordance with Expanded Greek-Hebrew Dictionary, electronic version (Biblesoft, Inc. and International Bible Translators, Inc., 1994, 2003, 2006), *faith*.
[2] "Now faith is the substance of things hoped for, the evidence of things not seen" (Hebrews 11:1 NKJV™).
[3] Hebrews 11:3b NKJV™.

the invisible, subatomic world: By reorganizing at the subatomic level we change what appears at the physical level.

These over-simplifications may help us grasp the idea. Microwaves that alter the state of raw food is an example of the invisible acting upon the physical and altering its structure. What about a radiation leak? It has the ability to change the tissue of our bodies or even alter the physical makeup of unborn babies. It can be something as simple as the rays of the sun burning our skin. These all are examples of the unseen altering the seen! Because these fit into the realm of our common experience we accept them without question or comparison to other levels of this same phenomenon.

I realize that all of these examples present the effects of something unseen that we can now measure. But remember, the fact that we can measure it makes it no less a part of the unseen world. These simple examples give rise to the reality that an exchange between the seen and the unseen is always occurring. At some point in time these interactions were considered miraculous. Because they occur in the realm of what we can now measure and explain, what was once accepted by faith is no longer faith. But as surely as these realms interact and bring about physical change, so the energies between the human faith and consciousness influence the unseen world, which ultimately alters the physical world.

An exchange between the seen and the unseen is always occurring.

In the book *Infinite Mind* Valerie Hunt does an incredible job of discussing the exchange between human consciousness and the quantum field. Quoting from Jahn and Dunne's book *Margins of Reality: The Role of Consciousness in the Physical World*, she presents some conclusions of great scientific minds concerning the interaction of mankind with the invisible world.

> Consciousness may insert information into its environment as well as extract information from it. The shadow nature of physics' reality paradoxically leads many scientists toward metaphysics or a mystical view of the world, like so many of our pioneer thinkers: Bohm, Einstein, Heisenberg, Jung, Penfield, Eccles, Eddington and Schroedinger. Jahn and Dunne have quoted Charles' Third Law, "any sufficiently advanced technology is indistinguishable from magic." In Science we do not explain things away, but we do get closer to the mystery. (Lewin).[4]

The growing body of evidence has plunged the scientific community into the realm of faith—a realm that may not be as foreign to them as they had once thought. In fact, the skeptics of faith may unwittingly be the greatest proof of faith. How? Faith conceptualizes in the mind what cannot be seen with the eyes until it can be experienced in the physical world. Everyone operates the law of faith when they take action based on the deep belief or persuasion that something is true. Every scientific breakthrough was an act of faith on the part of some great thinker.

Pioneering research and discovery is driven by a belief or deep persuasion. Yes, these skeptics may have seen something with their eyes or measured something with an instrument, but at some stage of research, an infinitesimal amount of scientific evidence gave rise to a

---

4   B. J. Dunne and Robert G. Jahn, *Margins of Reality: The Role of Consciousness in the Physical World* (San Diego, California: Harcourt Brace, 1987), 204 and Roger Lewin, *Complexity: Life at the Edge of Chaos* (New York: Macmillan Pub. Co., 1992), 133, as quoted in Valerie V. Hunt, *Infinite Mind: Science of the Human Vibrations of Consciousness* (Malibu, California: Malibu Publishing Company, 1996), 57.

belief that, at that moment, existed only in the mind of the researcher. That is faith. It is not religious; it is not about God; it is about our capacity to grasp and conceptualize something that cannot be seen or proven at that particular moment.

...Conviction and Evidence

Faith is a conviction. A conviction, in a court of law, is a verdict that is reached based on the evidence presented. That evidence may not provide one hundred percent proof. There may be evidence to the contrary, but there is enough proof to reach a verdict or conclusion. Likewise, our scientist who gets a few bits of data that point to a possibility reaches a mental verdict of a particular idea. On the belief that his idea is true, he forges ahead. This is no different from the person who gathers enough data to believe he can alter his destiny or health. Both have some evidence that the outcome they desire is possible. They may ignore overwhelming evidence to the contrary because they accept that what they do know—however lacking—is adequate proof to ignore what can be presently seen.

Faith is the evidence of things not yet seen. Some of the greatest philosophical minds of our time and of recent history believe this truth: The fact that we can conceive it is the proof it is possible. In the world ruled by the quantum field, all things exist in the realm of possibility. Einstein believed in a similar concept. Parallel possibilities—every conceivable outcome to any scenario—already exists. We choose the end of our situation by our beliefs and expectations.

The fact that we can conceive it is the proof it is possible.

The placebo effect is an unexplainable phenomenon within the scope of the scientific process. Drug studies conduct what is called a double-blind study for new drugs. Half of the patients are given the actual drug; half are given something with no known healing properties, such as a sugar pill, an injection of a saline solution or a surgery in which no organs are actually treated.

The placebo effect occurs when a patient who received nothing that should make him improve actually gets well or has an improvement in symptoms. Although studies vary, and some show higher numbers, a 1955 study by H. K. Beecher reported that thirty percent of the people who received no treatment actually got better.[5] These numbers are often greater than those of people who actually take the medication.

The word *placebo*, according to Gregg Braden in his enlightening book, *The Spontaneous Healing of Belief*, comes from a Latin word used in the beginning of Psalm 116:9 that translates as, "I will, or I shall."[6] It seems that the effect of the placebo is the result of the will of the patient.

Every day medical doctors in experiments around the world see people get well simply because they choose to believe they will. They believe the medicine they are taking will make them well, when in fact they aren't even taking medicine.

Acting on evidence that isn't true, they develop a conviction or persuasion that they will be healed. If it is not the medicine/placebo that gets them well, what is it that actually happens? Simply this: An interaction between consciousness and the unseen, quantum field brings about a change that alters the function of the very cells of their bodies. Trusting the wrong things, having completely erroneous evidence—whether religious belief is absent or present—they operate the law of faith. They become deeply convinced!

---

5   H. K. Beecher, "The Powerful Placebo," *Journal of American Medical Association*, vol. 159 (no. 17) (1955): 1602-1606.
6   Gregg Braden, *The Spontaneous Healing of Belief: Shattering the Paradigm of False Limits* (Carlsbad, California: Hay House, 2008), 42.

The Law of Faith

This realization of mankind's ability to alter the unseen world became a scientific reality with the advent of quantum mechanics. One of the most famous pioneers of this field was Werner Heisenberg, who transformed physics when he developed these radically new concepts in 1927. In his groundbreaking work *Physics and Philosophy* he states, "Natural science does not simply describe and explain nature: it is a part of the interplay between nature and ourselves."[7]

His insights lead us to realize that we never simply observe nature itself; we observe nature exposed to our method of questioning. We live in a participatory universe where all reality, or *matter*, is altered by the view, opinions and beliefs of the observer. This suggests that to some degree the world, including our physical bodies, becomes what we expect it to become.

The laws of quantum physics and interaction make the statements of Jesus, the Great Healer, understandable: "All things are possible to him who believes!"[8] Reality becomes what we believe it to be, whether good or bad, whether based on a lie or the truth. We choose our outcomes because we all operate the law of faith continuously. We all are persuaded of things. We act on those beliefs, but more importantly those beliefs act on the physical world to affect our health, happiness and prosperity.

I would like to expand the concepts of faith by adding this one idea. Faith trusts the laws of the universe; therefore, it not only believes there is an unseen quantum field, but it also believes it can act on that world by choice and expect the outcome of its choosing! In this concept is the first and most important key to entering and living in the realm of the miraculous: *All things* are possible to us because we know and use the power of believing!

---

[7] Werner Heisenberg, *Physics and Philosophy*, 1958. George Allen and Unwin Edition, 1959, as viewed on the Athenaeum Library of Philosophy <<evans-experientialism.freewebspace.com/Heisenberg.htm>> December 4, 2008.
[8] See Mark 9:23.

Faith then is not a determination of how much we have. It may or may not be connected to a religious belief. Faith is what we believe about the unseen world, our ability to interact with that world and its ability to alter the physical world inside and outside of our bodies.

Chapter 7

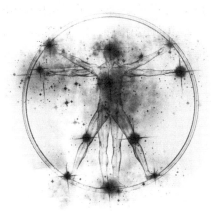

Crossing the Void

The fact that there is a physical world and an unseen world goes without saying. Yet, for any number of reasons man is aware of the physical world and has virtually no conscious awareness of the unseen world. The five senses associated with our ability to perceive and experience the physical world dominate every aspect of our lives and our life choices. The unseen, subatomic world is now known intellectually and explained scientifically, but this information has done little to motivate the masses to developing the *internal senses* essential for conscious awareness of the unseen. There exists in the mind of man a void, an unknown chasm that must be crossed to bring about deliberate, positive interaction between the physical and non-physical realms.

Man is stuck in a state of "default unawareness." Unawareness and ignorance are the default choices of the person who does not proactively choose to learn and grow. This default unawareness delivers us from accountability for our current condition. It keeps us from having to consider what can be difficult options, but more than anything else it frees us from personal responsibility. We didn't choose a limited life; we simply didn't choose a better life. But the odd thing about ignorance is while it provides a great excuse for our condition, it does nothing to improve that condition.

Barriers to Change: Fear and Pride

Man's willingness to live a limited quality of life is driven primarily by fear and pride. Discovering things that might demand too much change would place us squarely in confrontation with what many believe to be our greatest resistance to personal growth! If man learns what he does not currently know, he could be forced to rethink his every concept of life. This perceivably would be the ultimate destruction of the human race. *Perceivable* is the key word in that thought.

Perception of pain and real pain are not the same. They can, however, have the same effect! The idea that we might discover what would cause us to have to change how we function in this world would create a massive amount of pain in the form of effort and learning. But even more unnerving is the idea that we could be wrong about so many things. Apart from a strong sense of self-worth, man cannot face being wrong without enormous self-destruction!

Man's resistance to change is rooted in the fear of being wrong and the need to be right—that is, pride! There is a false security in the feeling of being right. Man really doesn't have to be right to feel safe; he only needs for it to *seem* he is right. As an ancient Hebrew proverb says, "There is a way that seems right to a man, but its end is the way of death."[1] In their insecurity the masses would rather be wrong and feel right, than to be right and have to work through and conquer the feelings of being wrong.

Insecure people live from a cultural conscience. The feeling of right and wrong emerge from what the people in our culture consider to be right and wrong. *To leave behind what is considered normal amidst social*

---

1   Proverbs 16:25 NKJV™.

convention and to create a new mind requires being an individual.[2] In the absence of a healthy sense of self-worth, no man functions as an individual. Rather, he is ruled by his culture, the group. The safety and acceptance and subsequent shared identity of the group are the substitute for self-worth and individuality. Losing the acceptance of the group is more than what the average person can endure.

We would rather be wrong and feel right than to be right and conquer the feeling of being wrong!

Pride is known to be totally destructive by every religion and every culture. The Bible says that pride precedes and produces destruction.[3] It seems that pride has the power to produce the very situations we fear while preventing us from the realities that we hitherto only have pretended to experience and enjoy. Dr. David Hawkins presents the possibility that pride is the last great vestige of resistance to change. In identifying the role of pride in resistance to change he says, "…radical change…can be disorienting; the courage to endure the temporary discomfort of growth is required, and the mind tends to resist change as a matter of pride."[4] The result is that man rather would live in his current pain than face the greater perceived pain of being wrong, thereby injuring his misplaced pride.

The unseen world is an unknown. For most of humanity the unknown represents a threat. A threat then produces fear, and fear represents the potential for pain. Man has an aversion to pain and will avoid perceived

---

2   Joe Dispenza, *Evolve Your Brain: The Science of Changing Your Mind* (Deerfield Beach, Florida: Health Communications Inc., 2007), 14.
3   See Proverbs 16:18.
4   David R. Hawkins, M.D., Ph.D., *Power vs. Force: The Hidden Determinants of Human Behavior* (Carlsbad, California: Hay House Inc., 2002), 220.

pain at all costs, unless the possibility of pleasure is great enough to justify the pain. Man has little emotional capacity for pain. The fight or flight mechanism is hard-wired into the nervous system as a way to respond to pain. We can run from it or fight against it, but we cannot passively endure it. As long as we are exposed to pain we will remain in the fight or flight mode, rendering endless damage to our physical and emotional health!

The problem with pain revolves around the idea of perception. Many things we perceive as pain really are not. Everything we need to make our lives better potentially could be perceived as pain. That's why people do nothing to improve their quality of life; the steps to make their lives healthier, happier, more prosperous and more filled with love are considered painful. And since we have never experienced those things at the level we desire, we are not even sure they are possible for us. So the pain seems to be a sure thing. The pleasure seems unlikely; therefore, the risk of perceived pain is not worth the remote possibility of the pleasure.

Stuck in Place

In his groundbreaking research and subsequent book, Dr. John Kappas explains how the mind processes information. He explains how thoughts enter the conscious mind and are passed to a part of the mind called the "critical conscious."[5] The critical conscious will evaluate information to determine if one should embrace or reject it. It is a well-established fact that the mind takes in thousands of bits of information per second and rejects all but four to six bits of that information. Dr. Kappas' theory helps us understand the process whereby information is accepted or rejected, as well as why we reject certain types of information.

In the critical conscious all information is evaluated in light of all we have known or experienced up to this point. In other words, does it

---

5   John G. Kappas, Ph.D., *Professional Hypnotism Manual: Introducing Physical and Emotional Suggestibility and Sexuality* (Tarzana, California: Panorama Publishing Company, 2001), 11.

fit into the realm of the known? If not, it is by default an unknown. And remember, an unknown without some very specific influence is rejected as a potential for fear and pain. This is why most people are stuck. They have no experience with the tools that empower them to move from the *known* to the *unknown* in a positive, healthy manner. They have no safe way to cross the void! At the same time, most of these tools are unknown. They are not taught in our *known* realms of education. As a result, they too are considered threatening and potentially painful.

Our way of thinking and doing things has become hard-wired into our brains. They are our automatic pilot. We do things the way we do with no conscious thought of how they affect us. We fear what our culture fears. Our sense of normal is defined within the scope of the thinking of the few people who influenced our lives. We are not only stuck, but we are stuck and don't know it! We are stuck and think we are free. We are stuck and the challenge of breaking free is far more threatening than the deception of stagnation.

In his book, *Evolve Your Brain*, Joe Dispenza discusses neuroplasticity, which at a very fundamental level is the brain's ability to adapt and create new habitual programming. In recognizing and maintaining the value of what we have learned in our culture, we must use our past as a launching pad instead of a docking station. It must be that upon which we stand to see the future, not that under which we hide to preserve the past.

Dispenza says:

To evolve, then, is to break the genetic habits we are prone to, and to use what we learned as a species as only a platform to stand on, from which to advance further. ...change is inconvenient for any creature unless it is seen as a necessity. To relinquish the old and embrace the new is a big risk. When we stop upgrading the brain with new information it becomes hardwired, riddled with automatic programs that no longer support evolution.[6]

Continual growth, learning and personal development are essential for the person who desires to free himself from the stuck state of the known and to cross the void into the unknown. In the famous parable of the sower and the seed,[7] Jesus presented a metaphoric concept about personal growth. In so many words He explained that the ability to grow and change is directly related to our ability and willingness to process information (seed) until it takes root in the soil (our hearts). Very true to modern discovery, He points out that information that does not take root can be lost from our consciousness very quickly.

> *Continual growth, learning and personal development are essential to crossing the void into the unknown.*

As profound as this principle of taking information into the heart may be, the statements that follow are equally sobering. First Jesus explained that the process whereby seed (information) takes root in our hearts (soil) is by consideration, thought, study and meditation. But then He makes the most shocking statement: Whoever has will get more, and whoever lacks will lose (or get more of) what he has (lack).[8]

---

6   Dispenza, *Evolve Your Brain*, 13.
7   See Mark 4:3-9.
8   See Mark 4:25.

Are You Willing?

People get locked into a way of living that exponentially grows. People who do not take care of their health follow that path until they actively destroy their health. People who take care of themselves keep finding ways to improve their health. People who prosper financially keep prospering, while those who don't do well seem to keep making more bad financial decisions and having more "bad luck." People who are "unlucky in love" move from being a victim to actively seeking out the very people with whom they can never find happiness.

Regardless of the terminology employed to describe this plight, it is true. In a program I call Heart Physics, I have developed the idea that the laws of physics that operate in the natural world are mirrored in the heart. They give us working models of how to bring about change in our hearts.

For example, the laws of physics state that a moving body will stay on a particular path unless acted upon by an outside body of greater force. People today are living in such an inescapable state. We are surrounded by people who think like we think. They read the same authors, attend the same churches or religious affiliations and even participate in the same recreation. We seldom develop friends of higher educational achievement. The life circumstances we have created produce an emotional reaction in us that reinforces the absoluteness of our perceived reality. Everything about the life we live reinforces the life we live. Because of our fear of learning something new we don't acquire the tools that would act upon our current state with such force as to propel us from our current emotional trajectory. We are picking up speed at a frightening pace that is hastening our lives into much, much more of what they have been up to this point!

You must decide if you want "better" enough to cross the void into quantum change. I am not asking you to leap blindly into the void of the unknown. I am asking you to be willing—willing to learn what you haven't known, willing to consider new possibilities and willing to put new actions into place. But before you do all that, I simply want you to be willing to peer into the void. Discover the way of painless, positive, permanent transformation.[9]

Crossing the void isn't really about the known and the unknown. Crossing the void is about the physical and the non-physical. The void that must be crossed is simply this: As children of the Creator, as citizens of planet Earth, we have the ability to influence the unseen (non-physical) in a way that changes the seen (physical) world. Accepting this reality opens our hearts and minds to endless possibilities!

---

[9] For more information, go to www.heartphysics.com.

Chapter 8

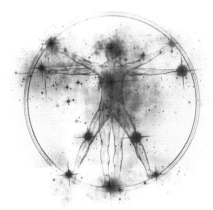

Shaping Our Possibilities

Our view of the world has been subtly but ever so effectively shaped by the sum total of our life experiences, including but not limited to everything we were ever told and everything we believed and assumed. Everything in our environment becomes part of our life experience. What we saw and even what we did not see are part of the accepted body of information that makes up our sense of what is possible or impossible. If we have seen it done, it is possible. If we have not seen it done, it is impossible—at least until new information moves it to the realm of possibility.

Our subjective experiences and interpretations forge our concepts of the possible and impossible. We established our sense of normal on what we saw and experienced. Once anything becomes normal, the standard is set—as well as the barrier. The barrier is the end of how far we believe something can go before reaching what we consider to be impossible. Barriers, however, are rarely real; they are perceived. They are the product of beliefs. A Hebrew proverb says that barriers or boundaries are a product of the heart's beliefs.[1]

---

1   See Proverbs 4:23. According to *Strong's Concordance*, the word *issues* could be translated as "boundaries." (Biblesoft's New Exhaustive Strong's Numbers and Concordance with Expanded Greek-Hebrew Dictionary, electronic version [Biblesoft, Inc., and International Bible Translators, Inc., 1994, 2003, 2006], *issues*.)

History repeatedly has attempted to teach us the lesson about boundaries, but we seem to keep missing the point. After millions of shattered barriers, we still believe the remaining barriers to be real. Prior to breaking the barrier of the four-minute mile for a runner, it was considered to be impossible. Yet, once it occurred, once the mental barrier to possibility was broken, it became almost commonplace. The same is true of inventions like the airplane, the telephone and the automobile. Once someone broke the barrier, technology zoomed forward at a frightening pace. In my mother's lifetime, men went from riding horses to walking on the moon—all because the barriers were being moved continually upward!

What Are Your Limits?

I once read an interesting story about an Olympic weight lifter who could not cross the five hundred-pound barrier for weight lifting. Time after time he would get ridiculously close to the five hundred-pound mark, only to fall short when attempting to break his record. In a competition, without telling him, his coach loaded extra weight onto his bar. Only after winning the competition and breaking a world record did the weight lifter discover he had crossed the five hundred-pound mark. The boundary was not in his muscles. Five hundred pounds was not the limit of his strength. It was the limit of his belief. It was beyond the realm of possibility in his own heart and mind!

Each of us is faced with our own self-imposed limitations and boundaries every day in every area of life! We are as happy and healthy as we believe we can be. We are as prosperous and successful as is acceptable to our personal beliefs. Our idea of what is possible, however, has been forged by the progress of our culture. It is as if there is a collective ceiling on how high we can go and how great we can be—a standard that is accepted and protected by all, at all costs. This was brought home to me so clearly when I read of a remote village where the inhabitants built their huts in a particular shape. If anyone built a hut in a different shape, the rest of the village would attack his family in the night, killing them and destroying the hut. As someone has so aptly stated, "Civilization has as its primary goal to hinder progress."[2]

We are as happy and healthy as we believe we can be.

Once normal is established (by the silent consensus of the group), there is an unspoken conspiracy to hinder and stop, if possible, anyone who pursues a higher goal. It seems that society feels threatened by those pioneers who have gone beyond the realms of perceived possibility. Men were murdered because they were willing to consider the world was round. Some of the greatest scientific discoveries had to be concealed, halting all of civilization's progress for hundreds of years, for fear of persecution or death. Millions of scientific and historical books were destroyed simply because they were written by people of a different religion or culture. To this very day, medical cures are rejected solely on the basis of cultural prejudice.

Where are the boundaries in your life? Who set the limits of possibility for you? Are you happy with the limits your life has reached? Would you expand those boundaries if you knew it could be done without pain? Almost every form of psychological research and therapy shows that people seldom rise above their social setting. Psychologists teach the value of re-socialization. The New Testament warns of the ethical, spiritual and moral influences of the people with whom we choose to fellowship. Few people will ever rise above the standards of their peers. Why? It is because

---

[2] Unknown.

our social group is our culture and our culture influences our sense of right and wrong, good and bad, acceptable and unacceptable and last, but by no means least, possible and impossible.

Perhaps the overuse of our five senses and the under-use of our inner senses is the ultimate culprit in establishing our limited sense of boundaries. We can know only what fits into the realm of our personal experience. Because we have interpreted the world from the realm of the physical for so long, we just don't realize the vast array of options out there.

Newtonian physics has ruled our world. Newtonian physics is actually very accurate for understanding physical matter. But with the advent of quantum physics we are now peering into the subatomic world that has, up until now, remained invisible and unknown. We now have rules for understanding how things work in this invisible realm. However, like all linear thinkers, people struggle with which one is right and which one is wrong: Newtonian or quantum?

There is really no need to choose one over the other. There is, however, a need to understand the benefits of both. We perceive the world through what have become our most highly developed senses. But these five senses only can show us the Newtonian manifestation of an even greater reality. Thinking that this physical world is only what we see is equivalent to thinking that our flipping a switch is the reason a light bulb gives light. If that is our belief, when there is disruption in the electrical current, we will stand helplessly in the dark, flipping the light switch and never getting the desired result. It might be good to know how to both work on the switch and change the bulb (Newtonian) *and* work with the laws of electricity (quantum).

If our five senses become the basis for anything other than understanding how things work at the physical level, we create a closed information loop that repeatedly brings us back to the wrong conclusion while ignoring the greater body of information and possibility. Valerie Hunt points out, "...our primary reference to the world is our physical body, and from these bodily sensations we create a reality in which in which our body is constantly present."[3] We are present in and experiencing the world we perceive, not the world as it could potentially be. But the world we see is made up and created from the invisible world we do not see, the world not perceived by the physical senses. We must experience a greater reality than the one afforded to us by these limited senses if we are to enter the realm of the miraculous.

The Key to Unlocking Barriers

The law of faith says, "All things are possible to those who believe." This means we can expand our boundaries, tear down the limitations and go beyond the known to the extent we believe it to be possible. In their book *The Body Electric*, Robert Becker and Gary Seldon discuss the fact that some species can re-grow limbs. Such experiments open the discussion as to why humans, who have all the components that produced our original limbs, could not re-grow a lost limb. Given what we now know about cloning, the information to perfectly reproduce any needed organ or limb is present in every cell of our bodies. The research of the last hundred years is ever more pointing to the possibility that if we could find the missing key, *all things* truly would be possible!

Maybe the one missing key is simply belief or faith. The deeply felt conviction that something is possible may be the one and only limitation to living in that reality. Faith is not a question of "how much," as in, "How much do I have?" Faith is more of a matter of "what," as in, "What do I believe is possible?" So if we expand the boundaries of our information, we may come to expand the boundaries of our faith. Faith or persuasion can

---

[3] Valerie V. Hunt, *Infinite Mind: Science of the Human Vibrations of Consciousness* (Malibu, California: Malibu Publishing Company, 1996), 60.

come from many sources. But those sources may be outside of our current circle of influence. Maybe we should look to the arenas that provide the two most overwhelming set of beliefs—that is, persuasion. In her chapter titled, "Science and Thought: The Real World," Valerie Hunt begins by saying,

> Information from science and religion mold our beliefs about reality, what we consider true, what imagined and what carries the strongest emotional charges. These pervading beliefs direct our energies, creating life patterns.[4]

For the first time in history, at least in the arena of the miraculous, we are finding much common ground between science and religion. If we are willing to open our minds to a shared vocabulary, we might begin to find overwhelming evidence that makes the miraculous much more possible.

We have the built-in tools to change our beliefs at will!

But the greatest source of expanding our sense of possibility is not the intellectual mind. It is in fact the realm of what some might call the subconscious mind, the "other than conscious" mind, or what I refer to as the heart. Since it is easiest to believe what we have seen and experienced, we have been equipped with the built-in tools to change our beliefs at will! We don't have to have information that comes from study and intellectual pursuit. All we have to do is experience the reality we desire. We must see it with the eyes of the heart: the imagination!

---

4   Hunt, *Infinite Mind*, 37.

Shaping Our Possibilities

Nearly every religion in the world places the utmost of importance on meditation. Meditation in its simplest form is when we think, ponder, consider, imagine or visualize something until it is experienced as real. Multitudes of research from scientific and religious disciplines have reached a similar conclusion: The mind does not know the difference between reality and what is clearly imagined.

Golf pros and athletes from every arena have learned the value of seeing themselves make the shot before they act. Sometimes just thinking something through before we act creates a new sense of reality. We see possibilities as to how it could work. If in meditation we saw something as real—more real to us than what our five senses are showing us and affecting our emotions more than anything in our current field of experience—we could slip into that reality.

By employing this simple discipline, great thinkers like Einstein developed some of their greatest theories and experienced some of the most profound breakthroughs. People of all generations have defied death. Cripples have walked. Wars have ended. Physical laws have been defied. People who live in the limitations imposed by their lives do not have to define your outcome. Start to see beyond the realm of your experience and begin to experience the realm of your goals and desires. When they are more real to you than anything else, they will shift your reality! In the realm of the miraculous, the only limitation is the boundary of your imagination!

Chapter 9

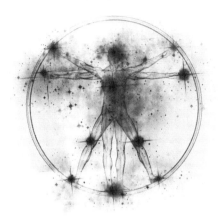

Shifting Our Awareness

At first glance miracles *seem* hard…no, they *seem* impossible! If we are holding on to incorrect concepts of the miraculous, then we rarely will ever experience a miracle and surely will never learn to live in a miraculous state. The old paradigm makes the miraculous seem unnatural. As we take on a more realistic view of Creation and how it operates, though, miracles become very natural. The old mystical paradigm led us to believe we must violate the very laws of nature to attain the miraculous. The new paradigm says we simply must harmonize with the immutable laws of nature and the miraculous will become "natur-al"!

Of all the paradigms that bind us to limited beliefs, the religious mindset may be the most limiting. The religious paradigm says, "I must convince God to violate the laws of nature for me. But before I can do that, I must meet all the religious qualifications necessary to believe that God will act in my behalf." Not only is this mindset limiting and destructive, but it also is inconsistent with the biblical model employed by the Great Miracle Worker: Jesus!

Regardless of our paradigms, the point from which we focus our gaze and make our assessment only looks at the physical world. Based on the Newtonian laws of physics, if we want to engender change, we must somehow generate a force or power greater than the obstacle we hope to move or change. Our ever-present connection with the visible, physical world shapes our every belief and defines our faith. So we somehow believe that we must generate a force of faith big enough to do the job. It is pretty daunting to think we must have enough faith to cure cancer, overcome devastating poverty or defy the laws of nature.

But remember, faith is a matter of what you believe and how deeply you are convinced of what you believe. Today you are arming yourself with the right information. You are discovering what to believe about the anatomy of a miracle. And you are being given the tools to persuade your heart, which can make anything believable. Now you're ready to engage the invisible and start the miraculous process!

"It's a Cinch by the Inch!"

For hundreds of years man used gunpowder to make explosives. In order to create more powerful explosions, he simply used more gunpowder. In time the explosives become more potent, but they still worked from the same premise: To have more power, use more explosives. This model was consistent with Newtonian concepts of physics. Bombs just got bigger and bigger.

Then one day someone looked at the issue from a completely different perspective. What if we looked at the smallest possible physical property? So began the atomic era. By splitting the atom, man created a bomb of such magnitude that we now have the power to completely destroy planet Earth. That massive power did not come from the physical realm ruled by Newtonian physics. It came from the invisible realm ruled by the laws of quantum physics.

In the same way we are not seeking to generate enough power to change anything on the physical plane; we are seeking to generate enough power to influence the energetic plane and set off the chain reaction that ultimately synergizes into limitless power. As Robert H. Schuller has said, "It's a cinch by the inch." Working from the smallest part is easy in any arena of life, but "it's hard by the yard." That which appears impossible becomes possible, in our hearts and minds, by looking at the smallest part!

Why is this so important? Easy! In order to operate the law of faith, we must make the miraculous believable to us! Jesus made this statement: "…if you have faith and do not doubt, you will not only do what was done to the fig tree, but also if you say to this mountain, 'Be removed and be cast into the sea,' it will be done."[1] At first glance this verse would seem to say that doubt is the absence of faith and that faith and doubt cannot exist at the same time. Upon closer examination of the original language, though, we see a much clearer picture.

The question is, "Which do I really believe?"

In the original language of the New Testament we do not have a concept of doubt *or* faith; it is not one *or* the other. Instead we have the concept of two coexisting, opposing forces or beliefs: "One thought pulling us in one direction; the other thought in the opposite direction…."[2] The question is not, "Do I have enough faith?" The question is, "Which do I really believe?"

For years I have used this simple analogy in my seminars. I would set a chair on the platform and ask someone to come from the audience. I

---

1 Matthew 21:21 NKJV™.
2 R.C.H. Lenski, *The Interpretation of St. Matthew's Gospel* (Minneapolis, Minnesota: Augsburg Publishing House, 1943), 824.

would have that person represent faith and the chair represent the problem he needed solved or the mountain needing moved. I would ask the person to move the chair. The person would easily slide the chair noisily across the stage!

Then I would interject an additional factor. I would say, "Now I will represent an opposing belief." I would hold to the chair and ask my participant to slide the chair across the floor. This time he couldn't. Why? Was it because the person didn't have enough faith? No! The person had the same faith (power) he had before. The difference was not the lack of faith; the difference was an opposing force or an opposing belief. But more specifically, the difference may be where a person places his attention.

We cannot escape the presence of the physical world, but we can escape the control and domination of the physical world. Consciousness—where we place our attention and what we are most aware of—determines whether we function in the Newtonian realm of the seen world or the quantum realm of the unseen.

Focus Your Awareness

Another great example is the "plank walking" exercise. As a young man I worked in construction at very dangerous heights. The first thing they tell you when you start working at these heights is, "Don't look down!" As simple as that advice may be, it is the best advice you can get. Looking down changes your awareness. That simple shift of attention will cause everything in you to change.

The plank walking exercise shows us a lot about focus. I often will lay a twelve-inch wide by twelve-feet long board on the floor and ask someone to come up and walk it. After the person has walked it several times, I ask, "Do you think you can walk that every time?" The answer is always a resounding, "Of course!"

"So," I continue, "if I put a one-hundred dollar bill on the other end of that board and all you've got to do is walk the board, you would do it?" "Absolutely," is always the answer! After a few more jibes, I add on an additional fact. "Okay, what I'm going to do is lay a one-hundred dollar bill on the board, then raise it fifty feet in the air. If you walk it, it's yours."

Now the person is not so sure. After walking the board a dozen times on the floor without so much as a wiggle, why wouldn't he take the offer? Simple! Now the individual's focus has changed from the reward at the end of the board to the possibility of danger. The person shifted his awareness from the reward to the possible pain!

The question now is, "Where is our awareness?"

The same thing happens when you attempt to believe for a miracle. Your awareness is directed by the five senses, not the heart. Our keen awareness of the physical world and our total lack of awareness of the invisible realm makes what we experience through our five senses more believable than what could be occurring in the non-physical realm! There is now an opposing force. We still have enough faith, but we are placing our attention on another possibility: doubt! To "walk by faith, not by sight"[3] doesn't mean we close our eyes and walk blindly through life. No! It means we walk with an awareness of what is more sure, more absolute, more convincing than what we can see, taste, hear, smell or feel. The physical world will not go away, but neither will the unseen world. The question now is, "Where is our awareness?"

---

3 See 2 Corinthians 5:7.

Becoming sure of the unseen will occur through various processes, but they all involve persuading our hearts and experiencing reality at an "other-than-conscious" level. Learning the truth about what happens at the subatomic level is very convincing. Meditating until we see and experience the desired end is incredibly convincing. Envisioning ourselves speaking and our words coming to pass is among the most powerful things we can do.

To continue this thought, in a parallel passage another writer of the New Testament added some of Jesus' additional teaching. This additional teaching may be some of the most misunderstood passages in the Christian faith, but it is some of the most empowering teaching about miracles. Jesus said, "Have faith in God. For assuredly, I say to you, whoever says to this mountain, 'Be removed and be cast into the sea,' and does not doubt in his heart, but believes that those things he says will be done, he will have whatever he says."[4]

Most scholars agree that He actually said, "Have the faith of God," or possibly, "Have the God kind of faith." It seems that God's faith operates through one very fundamental process. He sees and declares the end from the beginning.[5] And He is absolutely convinced that what He says comes to pass. God's process of faith doesn't seem to be anymore complex than that!

I know you're thinking, "Yes, but that's God." Remember, we are the offspring of the Creator. We are made in His image. We live in a world that cannot exist in the absence of intelligent life interaction. It stands to reason then that it came into being by intelligent interaction and initiative. This world came into existence by absolute laws; it is held together by absolute laws; and we have the capacity to operate within the scope of those absolute laws. The law of faith is absolute. If we see the desired end and are fully convinced that what we say comes to pass, then we declare the end and the end comes into being!

---

[4] Mark 11:22-23 NKJV™.
[5] See Isaiah 46:10.

Many religious assumptions are interjected into Jesus' teaching: things He did not say and things we add to fit our limited paradigm and further justify our limited lifestyle. He did not say, "If you speak to the mountain and believe, *God* will make it happen." He said, "If you are functioning in God's kind of faith, the immutable law of faith, and you speak to a mountain and tell it specifically what to do, if you do not doubt but believe, *it will happen!*"

A lifestyle of inner awareness is more powerful than hours of concentrated meditation in the face of an obstacle. The way of the miraculous is the way of inner awareness. Do everything you can to become more aware of the inner world than of the outer world. It is only as you move the mountains in the quantum world that you will move the mountains in this physical world!

Chapter 10

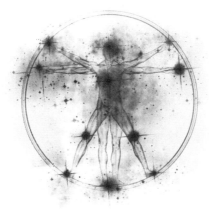

Quantum Power

Since faith means being deeply persuaded or fully convinced, then the more evidence we have that something *can be* and the more persuaded we become, the easier it is to have faith! Now that we have microscopes and other observational tools as well as measurements of movements and various activities of and for the subatomic level, it is getting easier and easier to believe in an unlimited capability! Having limitless faith is easier than ever before! Knowing what occurs at the invisible level is one thing. Knowing that we have the capacity to interact with and influence activity at that level is the first leap to releasing quantum power!

The tenants of the quantum era show that the experimenter, as a part of every research he performs, influences the results by what he chooses to study, the collection of data, and how he analyzes it – but even more importantly, by his actual presence in an experimental situation. It seems that scientists cannot extricate themselves from the experiment. They are part of the field of interaction and a basic ingredient of what they study. In this sense, truth on some levels is always relative.¹

From the initial discoveries about the quantum field to present research, the common consensus of an interactive field with intelligent life is undeniable! It does not require faith for mankind to influence the quantum realm; it happens just because we are here thinking and acting. Various experiments indicate that the state of world affairs at any given time is the combined beliefs and expectations of the inhabitants at that time, which sheds entirely new light on man having authority in planet Earth!

The uninformed scientist may believe the world to be on a random course of events independent, on the grander scale, of any significant influence from mankind. The irresponsibly religious may think that God is randomly making decisions about the course of world events. In their unawareness they are part of the combined field of beliefs that are guiding the events of the world.

The news broadcasters who love to capitalize on and sensationalize negativity are not just reporting the condition of the world; they are in fact influencing the condition of the world. The fear and expectations they induce become part of the field of shared beliefs. A Hebrew proverb teaches us that speaking positives or blessings build a city or nation, but speaking negatives destroy it.² Politicians who inspire fear and codependency as a means of winning votes, once having influenced the expectations of the masses, are actually programming our beliefs to bring about those fears. Whether it is good or bad, deliberate or by default, religious belief or scientific thought, when we deeply believe anything, we influence the unseen world to produce that belief!

---

[1]   Valerie V. Hunt, *Infinite Mind: Science of the Human Vibrations of Consciousness* (Malibu, California: Malibu Publishing Company, 1996), 43-44.
[2]   "By the blessing of the upright the city is exalted, but it is overthrown by the mouth of the wicked" (Proverbs 11:11 NKJV™).

We Are the Creators of Our Own Future

Man, just because he *is*, is a powerful force in the quantum field. He is always operating faith simply by having deeply felt beliefs. They may be chosen; they may be by default. They may not be what he desires. But they are all acts of faith that are molding his inner and outer world to conform to these very thoughts and feelings.

We think, when we feel a certain way about the future, that we are operating from some type of intellectual insight or a mystical gift. Although there may be elements of these factors, something far greater is at work. We think these insights are premonitions of events over which we have no control. We feel we are victims of the forces of the world around us. Then as these events unfold, we have what we believe to be evidence that supports our false theories.

Man is not a victim of the forces of the world. Neither is he a pawn in some cosmic chess game played by an unknowable Creator. Man is the creator of his world. We are not predicting the bad events that come into our lives when we have these fears, feelings and beliefs. We are *creating* them. The enlightened person who understands and utilizes the anatomy of the miraculous takes the feelings of fear and pessimism as a warning to reshape his beliefs, refocus his thoughts and choose a new ending!

Man is the creator of his world.

Renowned scientists and great thinkers are exploring the effects of consciousness on the quantum field. The results of those studies confirm what has been said by Taoists, Hindus, Jews, Christians, nearly every religion in the world and now the scientific community: Mankind has the capability to alter the seen, physical world by interacting with the unseen, energetic world. Each group's language may be different. Each may have added its own twists. The groups' individual explanations might not be accurate and their conclusions totally skewed, but each one is accurate in the shared belief of our ability to influence and change our inner and outer worlds.

We now know that matter is little more than an energetic field. At the root of all things is nothing—nothing physical, that is. Rather, the world is made of some form of energy. This energy can be acted upon by other energies. By introducing energy into a field we have the power to change that field. This process is demonstrated in an act as simple as boiling water. By subjecting the water to an outside force—heat—the water can change from liquid to vapor. Put the same water into a freezer and it will change from a liquid to ice. When any energy is introduced to any field, a change occurs in that field.

A misunderstanding of the science of thought leads us to view thoughts as non-physical, non-energetic, non-influential actions that have only limited subjective effect on the individual thinking those thoughts. Such an opinion ignores scientific research and thousands of years of empirical observation. Thoughts and feelings are the blueprints, the energetic encoding, for our lives. With our thoughts and feelings we influence the world around us; we alter the physical function of our bodies; and we even reprogram our cells.

Our Thoughts Are Electromagnetic

The physical world, at one level, is made up of atoms. We have some very clear understanding today of how atoms function. Gregg Braden in his book *The Spontaneous Healing of Belief* tells us that the old model of a mechanical atom, which looked like something orbiting something else, as in a solar system, has been replaced by a new energetic model. The new model looks more like circles of energy, resembling a round target instead of a solar system.

Braden says, "The important idea here is that energy in part is made in part of the electrical and magnetic fields – that create the thoughts of our brains and the beliefs of our hearts."[3] It is common knowledge that the brain emits energy. That energy can be measured by the use of an EEG. Just because these pads are stuck to our heads when measurements are made does not mean the energy stops when it reaches our skin.

When we think, our thoughts emit an electrical and magnetic energy into our bodies and into the world around us. Just the fact that electromagnetic energy is released clearly establishes the fact that we are influencing atoms. Our thoughts act on the very stuff from which all matter comes into existence and is held together!

But it doesn't stop there. If it did, it would just be random energetic influence with no way to choose our quality of life, bring about change or have any free choice. No, those thoughts are programmed; they are programmed with beliefs, expectations and feelings. That programming determines what type of effect specific thoughts render. It determines how those thoughts will affect our cells and our environment. Our every thought is charged with a program.

Ultimately we interact with and influence the unseen world regardless of whether we believe we do or we believe we don't. We will, to some degree, program a world that reflects our thoughts and feelings, both internally and externally. We all live in the miraculous. We all live by faith. But for those of us who choose the miraculous as a way to improve the quality of our lives and the world we live in, we must learn to be responsible and deliberate about our thoughts and feelings. We must accept the responsibility for the fact that the sum total of our thoughts and feelings are the ever-changing blueprint of our future. As such, we choose our future by choosing our thoughts!

---

3   Gregg Braden, *The Spontaneous Healing of Belief: Shattering the Paradigm of False Limits* (Carlsbad, California: Hay House, 2008), 58-59.

Suddenly, the miraculous is within your reach! You don't even have to try to change anything or anyone out there. Wallace Wattles says it so well in his book *The Science of Success*: "To set about getting rich in a scientific way, you do not try to apply your will power to anything outside yourself."[4] You are not trying to create a force. You are simply being who you are and doing what you do by choice instead of by default.

We choose our future by choosing our thoughts!

Choose the outcome you want for you (you can't include any other person in this), see the end, make it believable in your heart, speak the end, feel thankful and it is yours! Your thoughts and beliefs are charging the world in you and around you to create the end you think and believe! You always are creating this world, but now you have the ultimate secret to creating the world you desire: choice! Choose your thoughts, choose your feelings, and you are choosing your world!

---

[4] Wallace D. Wattles, *The Science of Success: The Secret to Getting What You Want* (New York: Sterling Publishing Co., Inc., 2007), 54.

Chapter 11

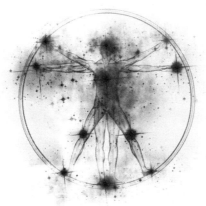

Sustaining Your Belief

I have experienced several incredible miracles in my life. Some were financial miracles. As an entrepreneur for more than forty years, I have seen many viable markets collapse. I have managed businesses through recessions. I have lost businesses and I have created businesses from nothing more than an idea. More times than I can count, it was applying these immutable laws that made the impossible a reality!

More than anything, I love the times when I see incredible, unexplainable changes in the circumstances of family and friends. Nothing moves me to compassion more than the needs of the people I love. And nothing warms my heart more than operating these laws and seeing the impossible happen for those people!

Probably nothing has improved the quality of my life and the capacity to pursue my destiny more than the times I experienced physical miracles. Having been born with a congenital kidney disorder that affected my immune system, I have faced and survived numerous life-threatening illnesses. I am alive because of the miraculous! Please do not misunderstand; many wonderful doctors and health care providers helped me stay alive as I persuaded my heart to believe in a quality of life beyond what they could offer. Without their help, I may not have lived to believe for the supernatural!

Regardless of the area of life we are dealing with, when we walk out the process of a miracle, the hardest part of the process is patience. *Patience* is the ability to endure under pressure without wavering. *Wavering* is when a person shifts from one opinion, thought or belief to another. We must remember that it is what we believe that programs our bodies and our environments. To shift from one belief to another is to create chaos both internally and externally.

Our ability to stand without wavering is based solely on how deeply convinced we are of the end we desire. Faith is a *deep conviction* or a *being sure* of the thing for which we hope! The more deeply we are persuaded, the less we waver. To the degree we are sure, we have patience!

When pursuing the miraculous, most people begin from inspiration. While reading this book you have, no doubt, been inspired to "go for it"! By this time you may have felt and lost inspiration many times. *Inspiration* is the spark that ignites the feeling of possibility. The word *inspiration* comes from the words *in Spirit*. The ancients considered inspiration to be God Himself sparking us to the belief that all things are possible.

Inspiration can be a spark or an explosion, but inspiration is not a belief. Turning an inspiration into a belief is a process. The majority of people who get inspired but do not see it through to the miraculous, think that inspiration alone is enough to get them through. But it isn't. It never is! Inspiration is the start, but faith is the process! Faith is the product of believing, but what is believing?

The Process of Believing

Because our five senses rule over us, we are easily persuaded that the physical is the greater reality. We, like the scientists who argue whether Newtonian or quantum physics is correct, are missing the point. We can't say one realm is real and the other is not. They are both real. The physical (Newtonian) realm is, however, the expression of the invisible (quantum) realm. The issue is not, "Which one is real?" The issue is, "In which realm do we seek to move, what laws apply and how do we perceive the desired end?"

When I felt pain in my body, it took very deliberate action to activate my inner senses and stay in touch with the end I had chosen. In the physical realm I was doing what needed to be done. I was taking medicine and operating in the laws of this realm. (In case you've wondered, taking medicine or doing anything else in the natural realm does not negate the effectiveness of the other realm. The key is to know which realm you're functioning in and use the appropriate laws!) But if I had lived by only what I could know through my five senses, I might have died. Instead, by keeping my heart "fixed" on me as a healthy, vibrant, active, happy person, I programmed the cells of my body to accomplish that end! As a result, *the end I had chosen stayed more real and persuasive than the pain I was currently experiencing.* I didn't waver! To overcome the domination of the five senses you and I must provide overwhelming evidence to the contrary. That only happens when we know how to activate the senses of the heart!

Remember, faith is not an intellectual persuasion. Intellect can be a part of it; information is certainly a part of it. But at the end of the day, our emotions always win over our information! Likewise, emotions always win over will power! *Emotions always win!* Therefore, unless we can direct our emotions, we cannot endure the conflict between the physical world and the unseen, energetic world![1]

---

[1] At this point, in order to avoid complications, I am deliberately using *emotions* and *feelings* interchangeably.

One of the greatest secrets of faith is this: Faith resides in the heart, not the mind! The heart is some combination of our souls and our spirits. It is the real us. It is where all that we are and all we have experienced culminate in the sum total of our lives. The heart is probably described in twenty-first century terminology by phrases like "subconscious mind." Regardless of the terminology, we know it is that part of us where we hold beliefs that supersede the intellectual mind. It is the autopilot. The thoughts and feelings that emerge from it drive us to actions that are often completely in conflict with our intellectual thoughts and intentions. Some researchers may be correct in their belief that this part of our being is linked to the sympathetic nervous system, which drives all of our bodily functions like heartbeat, respiration, cellular activity and all the other automatic functions.

> *Faith resides in the heart, not the mind.*

The thoughts of the mind can create specific emotions. As long as we hold those thoughts, we will sustain the corresponding emotions. The thoughts of the heart, however, are more about our abiding paradigms and beliefs. Therefore, the moment we stop putting forth effort to think a certain way, our thoughts drift back to our abiding beliefs—the beliefs of the heart! It is in this conflicted state that we lose the inspiration that made us believe we could experience the miraculous. When our emotions change, our beliefs change!

The truth is, beliefs and emotions go hand in hand. There is no separating beliefs and emotions. Beliefs and emotions are a continuum; one feeds into the other. What we believe creates emotions. Likewise, emotions, if allowed to continue, can alter our beliefs. It is essential that we harness our beliefs and emotions; this is the key to sustaining our beliefs and experiencing a miracle! To direct our thoughts and emotions is to direct our lives.

Beliefs and Emotions Go Together

Remember, faith is the evidence of what we can't see. Emotions are part of the non-intellectual evidence that forges our faith. Emotions can arise momentarily from any intellectual thought, but they disappear as quickly as another thought emerges. When a negative emotion arises, we should take that as a warning to initiate immediate counteraction. Instead, we may, intellectually, decide that the evidence is worth considering. This emotion or feeling can give us all manner of false evidence. We may feel we have an incurable disease. Based on that feeling we may search out all our symptoms on the internet. That information, mixed with the emotion, may cause us to conclude (believe) that we have an incurable disease. We may believe we have cancer. We may believe our death is imminent. Whatever we believe, we may be sure of this: We are programming our inner world and our outer world! Our bodies, minds and surroundings are doing everything possible to make things happen exactly as we *feel* they will!

The decision we must make concerning emotions is this: "Will I be led by emotions or will I lead my emotions?" One of my first encounters with this reality came when I had been symptom-free from my kidney disease for months. When I went in for a scheduled doctor's visit, I had a life-altering experience. I went into the office and walked to the receptionist window to sign in. I felt wonderful. As I stood talking to the receptionist, I smelled the "doctor office smell." I saw and heard all the things associated with my sickness. I began to have thoughts of sickness, and the fear that they would find something during this checkup began to surface. In a matter of minutes I started feeling bad. By the time I was called for a urine check, I had an infection. It happened that fast!

All those strong associations created feelings that led to thoughts that led to more feelings that brought about an almost immediate physiological change. Fortunately, while lying in the hospital I figured out what had happened and used the same process in reverse and walked completely out of that sickness! Those feelings presented me with evidence. The evidence was false, but I considered it as true. Now I never intellectually said, "These feelings are true." But I never did anything to stop the feelings, either. So they became real, more real than the belief that I was well! They gave rise to fear and doubt, or in other words, to belief in another outcome! To echo the words of Job, "The thing I feared came upon me."[2]

As we ponder the new *evidence*—whether intellectual or emotional—we sustain the negative emotion created by those thoughts. If we ponder that information long enough, we do what the ancients called "writing on the table of our hearts"![3] Pondering, considering or simply thinking until we create the corresponding emotion is meditation. Everyone meditates, whether one knows it or not or whether one believes in it or not!

Information plus emotion is how we write new beliefs on our hearts. Most of our heart beliefs were not chosen. They came prior to our fifth birthday and they guide our lives without our ever making a conscious choice about what we believe. While we were children words were spoken or thoughts were pondered in the presences of strong emotions, and those thoughts—however incorrect—became our core beliefs. Throughout our lives we write new beliefs on our hearts, but most of us do it by default and in conjunction with negative events!

---

[2] See Job 3:25.
[3] See Proverbs 3:3.

Jesus, the Great Miracle Worker, said, "...whoever *says* to this mountain, 'Be removed and be cast into the sea,' and *does not doubt in his heart*, but *believes* that those things he says will be done, he will have whatever he says."[4] The doubt to which He refers is nothing more than an opposing belief. It can be any belief that stands in opposition to the end you have chosen, believed and spoken. If this were merely a passing thought, it would come and go. You would return to the positive belief of your heart. But if a doubt, an opposing belief, is a heart belief, then it tends to abide. It is part of your automatic pilot. Unless that belief is changed at a heart level it will continually reemerge regardless of the amount of mental effort you employ to think positively.

Information plus emotion is how we write new beliefs on our hearts.

Verse 48 of the Gospel of Thomas may shed some clarity on what it takes to have a belief. It says, "If two (thought and emotion) make peace with each other in this one house, they will say to this mountain, 'move away' and it will move."[5] It is not just the thoughts we have that program our bodies and our worlds; it is the thought held in conjunction with emotion. Increasing the emotion is like increasing the electromagnetic influence on the subatomic world!

The only way to believe in the face of physical evidence is to feel the facts you have chosen. You must use your imagination to personalize your thoughts. You must imagine yourself living in the end you have chosen.

---

[4] Mark 11:23 NKJV™.
[5] "The Gospel of Thomas," translated and introduced by members of the Coptic Gnostic Library Project of the Institute for Antiquity and Christianity (Claremont, California) from the Nag Hammadi Library, James M. Robinson, ed. (San Francisco: Harper San Franciso, 1990), 137.

You must hold it and experience it to such a degree that you become fully persuaded of its reality! Feel the end result as powerfully as you would if it had just occurred. This is the power of meditation; it makes the future real today! When you can relax, see and feel the end that you have chosen, when it becomes your sense of reality, you can live in a new reality.

When the physical world gives you opposing evidence, go back to the place where you can see and feel the end result you have chosen. Continually feel well while believing you are well and you will probably stay well. Feel well while believing you are well when you are sick, and you will probably get well! Believe what you feel, and it will come into being! Choose what you feel and you are the master of your soul. Harmonize what you think and feel, and you are master of the miraculous!

Chapter 12

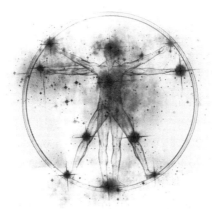

Programming Your World

Every person who feels victimized by the world wishes he had some control over circumstances. "If only I could have a say in the events that come to my life!" I have good news. That may not be the impossibility we once assumed. Based on scientific research, the Bible and dozens of other religious and philosophical concepts, having a say about our lives is exactly what is happening!

Our deep-seated beliefs may well be the program that shapes and drives the world around us, to some degree! The greatest proof of that reality is the very lives we live. Nothing testifies to the power of belief more than the quality of life we have today! It is impossible to separate our abiding thoughts and feelings from the quality of our lives.

An ancient proverb says that life and death are in the power of the tongue.[1] When that is coupled with the idea that words flow from the beliefs of the heart, we once again are faced with the idea that we are programming our world with our beliefs. It seems that the power of belief, however, is compounded by several factors. Each additional step we add

---

[1] See Proverbs 18:21.

to a thought causes a synergetic growth that at some point gains power exponentially.

The Steps Leading to Influence

In other words, we begin to create influence on the world around us by having thoughts. Then that influence, based on thoughts, gains power in proportion to how deeply we feel it. Its power is multiplied again when we act on those thoughts and feelings. To act on a thought or feeling is to operate laws in both the visible (physical) and invisible (energetic) world. Likewise, adding words to our thoughts and feelings influences others to have those same thoughts and feelings. The power of agreement—of others joining in our beliefs—can take something small to world-changing proportions on every level. But most importantly, all of these added steps serve to persuade us even more of the reality of our thoughts. And when it becomes a belief of the heart, it is no longer an occasional influence into our world but a constant influence in all we do.

> *The very opportunities that I attract or repel are those that match my sense of self.*

For example, if I feel unworthy, I interject that influence on every level. My words will subtly cause others to view me as unworthy. My actions will be driven by thoughts and fears of unworthiness. The very opportunities that I attract or repel will be those that match my sense of self. Every event that occurs around me, everything about my life, will be influenced to support my feelings of unworthiness.

The same is true of sickness, happiness, romance, prosperity and every other life pursuit. On both the physical and the energetic levels we influence the world around us to bring about our thoughts and feelings regardless of our desires.

The first thing a person usually says to me after discussion on this topic is, "Do you think I want these bad things to happen in my life?" Of course not! But because we are so in touch with our conscious thoughts and feelings and so out of touch with the thoughts and feelings of the heart, we seldom perceive the dichotomy that exists between our hearts and our minds.

The Power of Feelings

Wanting something is not the same as believing something! Desire is our hearts crying out for the manifestation of our needs! But we often want one thing and *feel* another! We want to prosper, but we don't feel there is any opportunity for us. We want our relationships to work, but at some very deep level we don't *feel* like they will. This is the two opposing forces of faith and doubt. All doubt is based in fear. Fear is the negative, pessimistic feeling that what we desire or need will not come to pass. At its deepest level fear is rooted in the rejection of self!

Whether the torment of pessimistic feelings or the rejection of self, fear produces very deep, abiding feelings and emotions. Remember, when there is conflict between a thought and a feeling, the feeling always prevails. When there is conflict between will power and a feeling, the feeling always reigns supreme. On every level, feelings always win. Wanting one thing and feeling something else may be the most tormenting existence possible. What is even more tormenting is to know one thing—to have the correct information—but to harbor feelings that keep us from acting on those feelings. This is the plight of the conflicted individual: disappointment, disillusionment, depression and despair!

Let's take a look at the power of feelings to program our world in contrast to the power of thoughts. This world we perceive as matter is made up of atoms. Atoms change when influenced by electrical or magnetic fields. We have established the fact that thoughts and beliefs produce electrical and magnetic energy. However, as Gregg Braden points out, the influence of the beliefs of the heart are far greater than those of the brain.

> Studies by the Institute of HeartMath have shown that the electrical strength of the heart's signal, measured by an electrocardiogram (EKG), is up to 60 times as great as the electrical signal from the human brain, measured by an electroencephalogram (EEG), while the heart's magnetic field is as much as 5,000 times stronger than that of the brain. What's important here is that either field has the power to change the energy of atoms, and we create both in our experience of belief.[2]

The beliefs of the heart are five thousand times more powerful at influencing and programming our world than our thoughts, wishes or desires. It is even possible that the more we desire something, the more we create the feeling of lack (not having it); therefore, the more we actually push it away from us.

Our society is entrenched in external change. External change places all the emphasis and effort in this realm, with little regard for changing the deep beliefs of the heart. That places all the focus on the problem and none on the solution. Externally, there is the ever-intimidating evidence of the size and magnitude of the problem. In this arena, if we get what we want or accomplish our goal, we can hold it only as long as we put forth the

---

2 Gregg Braden, *The Spontaneous Healing of Belief: Shattering the Paradigm of False Limits* (Carlsbad, California: Hay House, 2008), 59-60.

effort, will power and thoughts necessary to hold it. When we grow tired or weary, we can no longer hold it. We experience failure! After a sufficient amount of self-loathing and personal punishment, we try again. It is this very process that accounts for the repeated cycles of success and failure. On some level, everything we do on this plane makes it harder and harder to get and maintain what we truly desire.

The beliefs of the heart are five thousand times more powerful at influencing and programming our world.

Conflicting thoughts and beliefs make it nearly impossible for people to resolve the elusive riddle of their lives' struggle. Fortunately, when you seek to establish belief at a heart level, you don't need to know what is wrong. You need only to arm yourself with an absolute picture of what you desire, imagine it to be true and experience it as a reality. When the feelings of gratitude and satisfaction associated with the end result become an abiding reality, you have now harmonized thought and belief (feelings). All the powers of the world around you are now being aligned to make your desire into your physical reality!

Bring an absolute end to conflicting beliefs and emotions. Choose what you desire and think about it until you feel it as real, now. Do this daily, until it is the reality you effortlessly think and feel. While going about daily business, interrupt every conflicting thought and replace it with the thought you have chosen. Allow no internal conflict in thoughts and feelings! In so doing, you allow nothing to steal your future!

Chapter 13

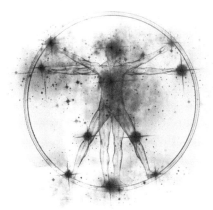

A Miracle in Your Inner World

One of the areas in which we can understand the anatomy of a miracle—an area with the greatest volume of scientific research—is in our inner world, in our physical and emotional health! Although mainstream medicine as a whole has not come around to thinking this way, many mainstream researchers and scientists are breaking through to what once was considered to be the realms of impossibility. These discoveries, while making spontaneous phenomenon understandable, also make it no less miraculous!

These cutting-edge, scientific breakthroughs are having an impact on nearly every area of scientific thought except health care. There are many roadblocks that slow the healthcare industry. The first is just the incredibly slow pace of research and integration. It is very difficult to test any medicine with absolute certainty of its safety or long-range effectiveness.

Regardless of the reason mainstream medicine is not moving fast enough, as long as we have our freedoms we can find people who are on the cutting edge of energy medicine and health care. Fortunately, there are no negative side effects involved in the process of walking in the miraculous.

Just like our outer world, the inner world is interactive and responsive to human intelligence, emotion and consciousness. At one time it was thought that our genes were the absolute program for our lives. We were told that our length of life, capacity for happiness and even the diseases that would probably take our lives were encoded in our genes and there was little we could do about it.

It seems that *scientific* concept was as reliable as other forms of divination and fortune-telling! The Darwinian philosophy of evolution correctly identifies but wrongly interprets man's capacity for survival and his dependency on learning. In order for there to be freedom of choice there must be the opportunity for bad choices as well as good choices. As the only species capable of improving our quality of life by our choices, the option for learning and growing must be a part of the Master plan.

We are *wired and equipped* with everything we need for an incredible quality of life, health, happiness and success. However, choice and type of programming is totally optional. As children we were programmed by the default mechanisms that gave rise to our current mental and physical dysfunction. As adults, however, the choices are now in our hands. If there is anything history has proven, it is the fact that any person from any walk of life can rise in influence, wealth and quality of life if that person has desire, drive and knowledge of the options.

Change Is Our Choice

Virtually every religious text in the world emphasizes the fact that we can change the quality of our lives. Although they all might disagree on the process, it is common knowledge among them that it is our choice! The oldest and most reliable scientific research in this field was traditional Chinese medicine. Thousands of years before our ability to do laboratory testing with human cells, the Chinese's empirical observation and whatever other forms of research they employed proved that we are not completely predestined by our genes to a fixed quality of health, happiness and life!

At the moment of conception we are genetically predisposed to certain factors. However, from the moment of conception until we choose to stop growing, we are able to reprogram the very cells of our bodies! Even the Bible makes an interesting statement about our health and our mind/emotions/feelings: "…prosper in all things and be in health, just as your soul prospers."[1] Or as it would be explained from a modern scientific perspective: Neural networks are the current understanding in neuroscience that explain how we change at a cellular level.[2] From every field of human health we see a consistent message emerging. We might not have any choice about how we start this life, but we have millions of options about how we will live it! Through the quality of our thoughts, emotions and beliefs, we create our health and our quality of life!

> *We might not have any choice about how we start this life, but we have millions of options about how we will live it!*

Traditional Chinese medicine (TCM) is used by more people than any form of medicine in the world! Many of the newest treatments in Western medicine are based on TCM. Of all forms of alternative medicine, TCM gets more consistent predictable results than anything I have ever witnessed. For example, for years Lincoln Hospital in the South Bronx of

---

1   3 John 2 NKJV™.
2   Joe Dispenza, *Evolve Your Brain: The Science of Changing Your Mind* (Deerfield Beach, Florida: Health Communications Inc., 2007), 48.

New York has used auricular acupuncture as it primary mode of treating substance abuse. This form of treatment, based on the scientific principles of TCM, has the highest cure rate of any substance abuse treatment I have ever heard about, read about or witnessed firsthand! The incredible success of TCM lies at least in part to its uncanny ability to effectively diagnose and treat illness based on the most outstanding concepts of body-mind connection on earth!

Western science is just beginning to grasp the body-mind connection in an applicable manner. TCM always has utilized the theory that our emotions affect our health and the function of our organs. In fact, TCM presents the idea that our health is affected by our emotions more than any other single factor! We know that our emotions are driven by our beliefs. So we no longer have to feel that our bodies are locked into a predetermined pathway over which we have no control. By changing our beliefs we can have our miracle in our bodies or emotions!

> ...DNA blueprints passed down through genes are not set in concrete at birth. Genes are not destiny! Environmental influences, including nutrition, stress and emotions, can modify those genes, without changing their basic blueprint.[3]

Learning to harness and proactively use this information gives us miraculous power within our own bodies and minds! Although the ancient sages knew nothing of the life of a cell, they knew they had influence in the unseen realm. Just as there is an interaction between our consciousness and the outer world, so there may be an even greater and more powerful connection between our consciousness and our inner world!

---

[3] Bruce Lipton, Ph.D., *The Biology of Belief: Unleashing the Power of Consciousness, Matter and Miracles* (Santa Rosa: Mountain of Love/Elite Books, 2005), 67.

The Role of Meditation

As we meditate and create deep, strong beliefs and feelings, we employ both left and right brain functions. It is the harmonizing of these two aspects of the brain—information (left) and emotion (right)—that influences our hearts and ultimately our bodies in miraculous ways!

Dispenza goes a step further and says, "...Meditation has shown promising results in changing not only how the brain works, by altering brain wave patterns, but also by growing new brain cells...."[4] Not only do we have the power to reprogram our cells, but we also can reprogram our brains to view life differently, expect a different outcome and manage our bodies in a way that can sustain the new outcome!

A cell's brain is in the outer membrane. There is research to support the idea that cells get their programming from outside sources, from environmental factors like thoughts and feelings. Cells have receptor antennas that pick up all manner of signals, including thoughts and emotions. It is these thoughts and emotions that give the cells the signals that direct our health, happiness and longevity.

Cells that start out controlling our lives based on the codes initiated by the combined programming of our parents at the time of conception are pliable; they await new programming to provide the life we desire *and* believe we will have. "A cell's life is controlled by the physical and energetic environment and not by its genes. It is a single cell's awareness of the environment, not its genes, that sets into motion the mechanisms of life."[5] When we, through beliefs, create a new environment for our cells, they will change to reflect that environment—whether good or bad; whether by choice or by default!

The moment something becomes real to us, we program our cells to facilitate that belief. They will give us the kind of health, sickness or miraculous recovery required to make our expectations a reality!

---

4   Dispenza, *Evolve Your Brain*, 58.
5   Lipton, *The Biology of Belief*, 15.

Spontaneous healings—that is, miracles—can occur when praying, when participating in a religious ceremony, when thinking about all the reasons we choose to live or when making a deep heartfelt decision. The moment something becomes a reality, our bodies produce chemicals more powerful than any drug on the street, our cells launch new programs and our outlooks and even the ways we think change! As Lipton continues to say,

> It is not gene-directed hormones and neurotransmitters that control our bodies and our minds; our beliefs control our bodies, our minds and thus our lives… Oh ye of little belief![6]

One source defines the Hebrew for the word *meditate* as "to plot."[7] To plot something is to lay it out or "to frame up"! In meditation we literally plot, lay out and frame up the future we desire. We make choices about our lives. At the same time we must remember that meditation is not what happens when we set time apart and withdraw from the world. We are meditating any time we ponder anything to the degree that we create the corresponding emotions, thereby framing up the end result.

Dr. John Kappas,[8] the man I believe to be the father of modern hypnosis, explained the process of one person influencing another into having a miraculous experience. It is no less miraculous even when we understand the scientific reason one person simply can command another person to "rise up and walk" and see it happen. Through the use of hypnotic language and several environmental influences, it is completely possible for anyone to inspire a miraculous occurrence in another person.

There are many ways we come to believe something. Repetition is one of the most common influences. As mentioned before, any situation

---

6   Lipton, *The Biology of Belief*, 28.
7   The Online Bible Thayer's Greek Lexicon and Brown Driver & Briggs Hebrew Lexicon (Ontario, Canada: Woodside Bible Fellowship, 1993), *meditate*. Licensed from the Institute for Creation Research.
8   To find out more about Dr. Kappas and the Hypnosis Motivation Institute, go to www.hypnosis.edu.

that introduces strong emotion and information at the same time, whether through meditation or the influence of a person to whom we make ourselves suggestible, have the power to set off a charge of miraculous processes in our inner world.

In my research and clinical experience, I personally have seen nearly every imaginable miracle. But what is even more astounding is that I have seen them occur in every imaginable way! When looking at them from the outside, many mystical conclusions can be drawn. But upon closer inspection, the anatomy of every miracle contains some very simple factors. The one common denominator is belief; the variable is the way the person arrived at his belief!

The one common denominator is belief!

By employing several methods at one time, we can have an influence on the physical and non-physical realms simultaneously, thereby multiplying the effectiveness of our efforts! By gathering new and positive information and creating positive thoughts, we can create one level of influence in our inner world. But by combining that level with meditation, prayer or some form that creates emotion, we can develop deep-seated beliefs. "Positive thoughts have a profound effect on behavior and genes, but only when they are in harmony with sub conscious programming."[9] If we truly want to hasten the process, we should include the conscious and subconscious as well as exercises, diets and other modalities that support the efforts of our inner workings.

---

9   Lipton, *The Biology of Belief*, 30.

The same principles that make it possible for your heart and mind to program the outer world and rearrange atoms occur in your very body. You have the power to create a miracle in your body. It is your right as a child of the Creator and as a citizen of planet Earth! You and you alone hold the keys to a miracle in your inner world. Propel yourself toward your miracle and gain exponential speed as you utilize every tool and employ every resource!

Chapter 14

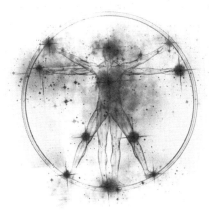

The Miraculous Power of Being Present

The world as it has come to be and the systems upon which it was built do not lend themselves to any of the inner workings of man. This world's system was built on the principles of force, not cooperation; on interference, not harmony. Maybe it is the Darwinian focus on survival or the slow but consistent loss of who we are in relation to the Creator; maybe it is just disbelief, but whatever the reason, there is little in an average day that facilitates the miraculous.

The law of harmony may be one of the most unknown, overlooked and misunderstood aspects of accessing miraculous power. In the linear, either-or world of the twenty-first century mind, there is passive and there is force. Neither of these options fare too well over the long haul. They both lead to incredible emotional conflict, resulting in stress, tension, conflict and fatigue. But thankfully there is a third option: harmony!

The Taoist philosophy of the ancient Chinese, like that of the Hebrews, sought to understand the *way* the world really worked. This understanding extended beyond the natural world to the emotional and physical. Instead of using force to violate the laws of nature, health and relationship, these ancients sought to identify them and cooperate. Instead of spending our lives swimming upstream, maybe it's best to let the stream take us where it is going!

This doesn't mean we passively let anything happen. Rather, as Einstein pointed out, all the possible endings to any situation exist in potential form. We have not chosen our end if we will not yield to the process required to reach that end. This could be one of the differences between choosing and wishing. An ancient proverb says, "Sensible people keep their eyes glued on wisdom, but a fool's eyes wander to the ends of the earth."[1] As this wise saying points out, both types of people may want the same thing. One looks only at the end for which they foolishly wish while the other looks at the wise (practical) way to get to the end.

Keep your eyes glued on wisdom!

Wisdom, the way to arrive at the desired end, is always present and is always calling out to lead us to the fulfillment of our desires. All people have to do is keep wisdom ever before their eyes and yield to it. To do so is to ensure the end we seek. To look wistfully at the end that seems so far away and wish it could be ours is to play the victim game! Wisdom chooses the desired end and then yields to the process to reach that end, thereby avoiding the stress of forcefully working against the natural process.

The way of force says, "I know what I want now based on all that I have experienced about life thus far; I will devise my plan and make it happen!" Consider this. If we are seeking what we do not have, we are seeking what we do not know how to get! If we had known how to get it, we would already have gotten it. No matter what the similarities are, even though we can apply the same principles in every situation, the subtle

---
[1] Proverbs 17:24 NLT.

nuances of every path require that we be present, paying attention and yielding to the process.

In every situation there is a paradox. Past successes and what we learned from those successes are an essential part of our process for current success. But when considering the *critical factors* for success, we must realize that paying attention and being present is a part of the process that cannot be set aside in favor of following a set of steps! We can draw from our past experiences, but let's treat every new venture with the special treatment of something new. And the main thing humans do with new things is pay attention.

Consciousness: Paying Attention

The study and practice of consciousness is as old as human beings. It was the way of the first inhabitants of planet Earth, and it has been the way of the wise for millennia. In my consulting work, I have found this aspect to be one of the greatest struggles for the up and coming leaders of the future. Great artists of every kind have the capacity to become so present, so conscious of their feelings in the present moment, that they are able to express those feelings through their paintings, music and writings. Great martial artists become so aware of themselves in their environment that they can fight with their eyes closed. They feel themselves and their opponents. Exceptional corporate leaders just have a "feel" for every aspect of the business plan they are working. Charismatic politicians seem to know just what to say and when to say it. Gifted ministers experience what they call the "flow." The one thing these people all have in common is the ability to be present in their situations and yield to the process!

Consciousness is how we interact with the world around us. Consciousness is how we influence our physical bodies. But consciousness is more than just awareness. It is an interaction with our inner and outer worlds that is ever aware of the role we play in the unfolding of events.

I define consciousness as our proactive, deliberate interaction with the unseen world. In his book *Evolve Your Brain*, Joe Dispenza gives a great concept of consciousness: "Consciousness enables us to think, and at the same time, to observe our thinking process."[2]

Consciousness is a proactive, deliberate interaction with the unseen world.

One ancient discipline requires the adherent to spend time every day becoming aware of each organ of the body. When the person would become aware of his liver, for example, he would notice how it feels. He would be thankful for his liver and smile into it. Doing that with each organ of the body would improve the health of that organ.

As a skeptic I remember thinking, *Yeah, right!* But as I studied the effect of consciousness, I was amazed at what I found. Where we place our attention actually does change the neurological and physiological activity of that organ. When we put our attention on a specific area of our bodies, the blood flow in that area changes and all manner of neurological activities occur.

Placing our attention on an area of pain actually makes the pain worse. Listen to this profound scientific reality:

> Our attention brings everything to life and makes real what was previously unnoticed or unreal.... according to neuroscience, placing

---

[2] Joe Dispenza, *Evolve Your Brain: The Science of Changing Your Mind* (Deerfield Beach, Florida: Health Communications Inc., 2007), 67.

our attention on pain in the body makes the pain exist, because the circuits of the brain that perceive pain become electrically activated. If we then put our full awareness on something other than pain, the brain circuits that process pain and bodily sensations can be literally turned off...[3]

This only describes a single process of which we are aware. How many *unknown* things happen when we shift our focus to become aware or to shift our awareness? The scientific study of consciousness is limited because of the current mindset. The dogma of the scientific world doesn't consider what it doesn't understand. The prejudice of ignorance is rampant in the area of consciousness. Because consciousness is relatively unknown, scientists traditionally have not considered even asking the questions that lead to discovery. The sensitive equipment needed to measure the effects of thoughts and consciousness has not even been made. The religious ignorance and resulting prejudices of a puritanical religious mindset in the West consider this biblically based activity of meditation to be taboo as well. Due to abounding opposition we have discovered only the tip of the iceberg.

But we know that our thoughts, beliefs and even our attention given to something creates enormous energy that influences the subatomic world around us, the cells within us, the way our brain works and an incredible shifting in neurological and physiological functions. Whether or not it can be explained scientifically or intuitively realized, we are *wired* to lead our lives by our choices. When we make a choice to be happy and spend time paying attention to it, we intuitively will discover the path that leads us to our choice. It is this very yielding to intuitive wisdom and violation of intellectual domination that is so threatening.

---

[3] Dispenza, *Evolve Your Brain*, 3.

Using the Consciousness

Several years ago I began to study a process called PhotoReading.[4] When PhotoReading, a person learns to take a mental photograph of the page while in a meditative state. The average speed is somewhere around two pages per second. Upon completion of the book, which is only a few minutes for a book of several hundred pages, the person makes some positive affirmations, closes the book and then waits about twenty-four hours to do some recall exercises.

When using this process to study for exams, I would simply sit at my desk, go to a meditative state and affirm that I possessed all the information needed for the test and that this information would come to me as needed. I would read the questions, follow my intuition and put what I felt the answer to be. Although I still to this day am taking courses, I have not studied for a test in ten years. And I have made As and Bs on every test I have taken.

But here is the hard part. Learning something new about how my brain works was great as long as it was no more than an intellectual pursuit. But when I started reading a book, upside down and backwards, at the pace of two pages per second,[5] that I would be tested on, I had my greatest challenge. I had to defy the first and most fundamental part of education. I had to trust that my brain really did work the way I had learned it worked and I had to harmonize with that process. Then when I took the tests, I once again had to trust that my intuitive wisdom was far more exact than my intellectual knowledge.

Awareness of the invisible world requires the development of internal senses long neglected by most people. But just as we came to rely on our outer senses, so we will sharpen and come to understand our inner sense by constant use. Like the scientist conducting experiments at the subatomic realm, we come to understand that our very observance of that

---

[4] If you want to find out more, go to www.learningstrategies.com.
[5] I do it this way when PhotoReading to avoid the temptation to try to read employing traditional methods.

realm alters the outcome of the experiment. In time we learn to use our faith, observation and awareness to create the miracle of our choosing!

Chapter 15

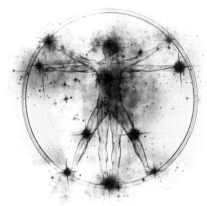

A Heart for the Miraculous

The great fallacy of miracles is that they are for the deserving. Miracles are not for the deserving; they are for the believing! Just as the laws of nature work for both the good and the bad, so do the laws of the miraculous. They are, after all, only higher laws of nature! They were given for our good. How we use them or whether we use them is our choice!

Although miracles are not just for the deserving, the law of the paradox applies again here. We don't have to deserve a miracle to get one. But if we do not *feel* deserving, we usually can't get one! It is true; hearts that are overrun with guilt and shame rarely can break through to the miraculous. Looking at this paradox from an external perspective causes one to reach erroneous religious, idealist and even legalistic beliefs. But by understanding the anatomy of a miracle, we know the absolute laws that must be present—and deserving is not one of them! It is here that almost every religious belief in the world drags us away from what is freely given for our good and takes us into a world of destructive introspection!

The Heart Has a "Memory Chip"

The heart is where we have our sense of self. We do not yet understand the degree of overlap between the physical heart and the spiritual heart, but we can learn about the spiritual by observing the characteristics of the physical. The physical heart is now known to store memories. Ancient Chinese medicine as well as the Bible tells us that the heart is the residence of our sense of self. Therefore, anything that affects our sense of self, affects our hearts and thereby our capacity for the miraculous.

Attempting to believe anything in our hearts always will be intertwined with our sense of self. Thus, the signal that emits from our hearts onto the world around us is not just what we believe about a miracle; it is overlaid with what we believe about us in relationship to that miracle. The two are inseparable.

Oriental medicine made the discovery—which is now scientifically disclosed—that the heart houses the long-term memories. This is why elderly people who have a strong heart can remember what happened fifty years ago but can't remember what happened yesterday. Actually, every organ in the body influences some aspect of our thoughts and emotions. But the heart affects our sense of self more than any other organ. As the residence of our long-term memory, it is the place where the sum total of all we have ever believed or done comes together in one collective sense of self!

Our hearts house our long-term memories.

In my clinical practice I have observed that people with heart issues have trouble with their own boundaries. As a result, they also have trouble recognizing the boundaries of others. This is incredibly destructive to relationships. This also helps us understand the concept of loving with our hearts. We can't really love others beyond the love we have for ourselves. Since love emerges from the heart, that love is colored by our sense of self. How we see ourselves dictates how we see others. We never can have a view of the world or of other people that is not defined by our sense of self. It is, after all, our sense of self that establishes our boundaries. And it is our sense of self that affects our interaction with the invisible world, the world of the miraculous.

Some research indicates that nothing is more destructive to our sense of self-worth than trying to become! Religious idealists wrongly would assume that our inability to experience a miracle is based on some deep personal flaw. They ignorantly would create a heart boundary. They would make us believe that we must discover and correct this flaw before we will qualify for our miracle. This concept is based in part on the idea that God must individually approve every request for a miracle. His answer would be subject to a moral evaluation that we would either pass or fail! This sets the unenlightened traveler on a quest to *become*. The need to become is precipitated by a rejection of self.

We all have flaws. If the miraculous is based on the flawless perfection of a moral inventory, then we all are in trouble. In fact, when speaking of people who made the miraculous occur, the New Testament says they were all men of like passions as we are.[1] It is not the one who believes in his perfection who obtains a miracle; it is the one who believes in these absolute laws and applies them.

"Garbage In, Garbage Out"

The heart is much like a computer. It runs programs and stores and retrieves data; the data may be correct or incorrect. It simply stores and programs our sense of self based on the evaluation of the data we had

---
[1] See James 5:17.

when we wrote it on our hearts! Like the hard drive on a computer, data is retrieved based on requests. Our hearts will give us the type of information we ask for. So when we ask, "Why can't I get my miracle?" then, with no bias or prejudice, our hearts answer the question the way we asked it. We perhaps meant to ask, "What do I need to do to get my miracle?" Nevertheless, the answer will not be based on our intention, but on our wording. Therefore, we begin to have thoughts and memories that answer the question and tell us everything that is wrong with us!

We mistakenly take this process to be the evidence that there is something so wrong about us that we are unqualified for a miracle. We now have created a boundary in our hearts. This boundary only exists in our minds and beliefs. Our attention and action are taken away from operating the irrefutable laws of the miraculous and now are seeking to solve the illusive problem. The more we see that is wrong with us, the more we are convinced of our lack of qualification and the more new boundaries we create. It is a never-ending cycle! It is rare that we ever reach the ideal moral inventory. Instead we reach the conclusion that God is dissatisfied with us—and the rest is a syndrome of personal destruction!

Then there is the problem of guilt and shame. Although a destructive life does not disqualify us, negative emotions affect our hearts in such a way that our cells and our world are programmed to give us the penalty we believe we deserve for the life we have lived! Guilt is a warning light; when we feel guilt we immediately should stop what we are doing. The real issue isn't whether it is right or wrong; the issue is, "How is it affecting my heart?"

Some studies indicate that people often have some of their greatest life crises at their moments of greatest opportunity. When considering all the factors, it may well be that the secret to this issue lies in the beliefs of the heart. If we don't feel deserving of our retirement, our hearts could program our cells to end our lives. If we feel unworthy of being loved, finding the love of our lives could cost us our lives! People who feel unworthy of the good things that come their way usually manage to destroy those good things or even themselves. A heart filled with guilt has little capacity to believe for a miracle, unless it adheres to the principles we have previously established.

> *A heart filled with guilt has little capacity to believe for a miracle.*

Shame is a whole greater level of negative feeling and emotion. While guilt looks at our action and rejects the action as being wrong, shame looks at ourselves and rejects us as a person. In his groundbreaking book *Power Versus Force*, Dr. David Hawkins identifies shame as the most destructive negative feeling a person can have, before death:

> …shame is perilously proximate to death, which may be chosen out of shame as conscious suicide or more subtly elected by failure to take steps to prolong life.[2]

Live your life with a clear conscience. Walk in love. Be a peacemaker. Don't allow wrath to fill your heart. When you have wronged another, do what it takes to make the situation right and find peace in your own heart. Accept the love of the Creator who gave you all these good things for your enjoyment. Keep your conscience clear of all things that would condemn you and make you feel disqualified. Realize that you exist in a world that is designed for healing, nurturing and an incredible quality of life. Trust the laws, not your goodness!

---

[2]   David R. Hawkins, M.D., Ph. D., *Power Versus Force: The Hidden Determinants of Human Behavior* (Carlsbad, California: Hay House Inc., 2002), 76.

Don't wait until you are in need of a miracle to attempt to develop your heart. Live a life where your heart is at peace, you are free from guilt and your words mean something to you. Remember, you have to see the desired end. So live a life that learns how to see the invisible. You have to speak to the situation and believe that your words come to pass. (Don't talk so irresponsibly that you have no value for your words.) Then believe that it is yours. Learn to live out of your heart in such a way that you direct your emotions from your ability to experience the end you desire. Develop a heart for the miraculous and having miracles will be your way of life!

Chapter 16

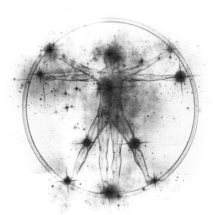

Entering the State

Too often we make the miracle a goal instead of a state. In fact, many of the things so essential to quality of life have fallen victim to the same faulty logic. When a state is confused with a goal, the entire process can become negative and self-destructive! A goal gives rise to thoughts of emergency, transition or expedience. A state addresses something habitual or abiding: a way of life!

Our quality of life is the product of our hearts' beliefs more than anything else. The problem, however abrupt it may seem to have appeared, was very probably a long time in the making! What we believe about ourselves and what we believe we deserve have been programming our inner and outer worlds. There is an ancient proverb that says, "The curse does not come without a cause."[1] The cause is always the beliefs of our hearts.

When I ask people if they feel they do not deserve to be happy and healthy, almost no one admits it. They're not lying; the problem is that they have lived with no awareness of their heart beliefs for so long they have no capacity to get in touch with their deep beliefs. On a conscious

---

1   Proverbs 26:2 BBE.

level, they have justified all their actions. They have found a way to make the choices of their lives palatable. They have figured out who to blame. Or maybe they have just forgotten how to hear their hearts. Whatever the reason, they have ignored the *voice* of their hearts for so long, they no longer recognize it!

Deal With Guilt; Don't Ignore It

When we violate our conscience, we feel it immediately, or at least as soon as the feelings of temporary pleasure have subsided. If we ignore the negative, stressful feelings of a violated conscience long enough, we stop feeling them, falsely assuming they have gone away. In order to function, our minds have the capacity to become insensitive to the feelings we ignore. It is part of our survival mechanism. But make no mistake; the consequences are still at work in us, and not just on a conscious level.

Ignoring guilt is like taking pain medication for a broken foot instead of setting the bone. We still can walk, but every step we take exacerbates the problem. When we do eventually get in touch with that pain, it may be beyond our capacity to resolve. Not feeling pain does not equate being well; it usually means we are just numb!

> *Ignoring guilt is like taking pain medication for a broken foot instead of setting the bone.*

Although we no longer feel the stress, guilt or negative emotion, it still affects every aspect of our subconscious. If our foot was broken and we continued to walk on it, our hip, spine and other parts of our bodies would begin to go out of alignment to compensate for the inadequacies of the foot. Our bodies would have to change the way they function to allow us to walk. While preserving our ability to walk, these compensations could create lifelong spinal problems. Likewise, adjustments made on a physical or emotional plane, designed to incorporate dysfunction, causes the entire emotional being to be out of balance!

We might be completely unaware of it, but compensations happen on the emotional level as well. We find ways to think, process, evaluate and see life that integrate the guilt as an acceptable norm. In time guilt feels like stress. Eventually we can't identify where the stress came from. The cause of our dysfunction will eventually become part of our life systems. Once we develop all these systems around the guilt, we then need that guilt to function. Eventually we resist releasing it. It has become such a part of who we consider ourselves to be that we wouldn't know how to function without it!

If allowed to remain unchecked, we eventually build an entire lifestyle that makes sense only when stress, guilt or shame is integrated into our identities. Now the kind of thoughts and feelings we need to harmonize with the laws of the miraculous violate our entire sense of self. We've incorporated guilt into our concepts of God. Our emotional energy is consumed with managing negative feelings. We don't know how to do anything but expect the worst. Our emotions become self-fulfilling prophecies!

Einstein said it like this: "We can't solve problems by using the same kind of thinking we used when we created them." A particular set of beliefs and a certain lifestyle played a role in creating our problems. It will take new beliefs and a new lifestyle to solve our problems. A temporary adjustment in thinking only brings a temporary solution. Unless there is a *change of heart*, we eventually will slide back into our old programming and recreate a similar circumstance or problem.

Goals Are Not States

For example, peace is a state, not a goal. To approach peace as a goal, people may attempt to remove as much conflict as possible. They may separate themselves from as many disagreeable situations as possible. Now while these are certainly worthwhile actions, within themselves they do not ensure peace. They may create calm. But the absence of conflict does not equate peace. Many people who have no conflict live in dread of conflict. Circumstances do little to give them the peace they work so hard to acquire!

The states that contribute to our quality of life are obtainable only internally! It is impossible to control enough of the outer circumstances to live in peace. Just the attempt to exert that much control will create conflict. Peace is an inward quality. It has to do with inner confidence. It is not a form of tranquility subject to external factors. Real peace can rule our hearts and minds even when we are surrounded by conflict. Circumstances cannot control peace!

The same is true of health. When there is health in people's hearts, they manage their lives for health. Their inner programming leads to healthy foods, healthy activities and an all-around healthy lifestyle! Individuals who have health in their hearts recognize the first signs of sickness and deal with it at an internal energetic level before it becomes physical. Or, at the first physical signs, they instinctively take steps that lead to immediate recovery!

Prosperity is also a state. Prosperity is not reflected by a person's capacity to make money. Anyone who gets a good education and a good job can make money. That may be no indicator of prosperity. People who have prosperity in their hearts always feel abundance. They enjoy what they have. They don't feel the need for more to be happy. People of true prosperity are generous. They feel a freedom and joy when giving to worthy causes. They feel they are contributors to the world. Likewise, they manage their money well. If they suffer a financial loss, they have no fear. They know they will prosper. They know they attract prosperity.

All of the things that truly contribute to quality of life are states, attitudes and inner qualities! They *only* can be obtained in the one realm where we have complete control and complete freedom of choice: our hearts! When something becomes a belief of the heart, the law of attraction works around the clock. Even when we are not consciously thinking about a certain goal, we are attracting and creating the expected end result by the abiding beliefs of the heart.

States, attitudes and inner qualities can be obtained only in our hearts.

One of the places I learned the power of entering a *state* was when I learned PhotoReading (which I talked about in Chapter 14). In traditional learning, which harmonizes with the lowest functions of the brain, we have to put forth effort to learn. With PhotoReading we enter the learning state and our minds effortlessly take in the information on the page at a rate inconceivable to the conscious mind! So if I enter a state of prosperity where prosperous is how I feel about me and how I feel about life…then I will naturally and effortlessly attract prosperity. I will find the good deals. I will be attracted to the decisions that produce prosperity. The same is true of health, love, happiness, peace and every other quality inherent to an incredible life!

The miracle of health, success, peace, happiness, love or prosperity is no longer quenched in the unrealistic demands of the illusive pursuit of extravagant faith. I do not need enough faith to make these things occur. I do not need enough faith to make things come to me. I do not need enough faith to persuade God to give them to me. I simply need enough faith to enter the state where all these things exist!

Chapter 17

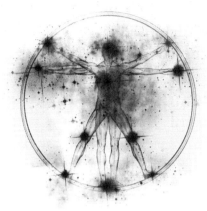

Releasing

Carl Nightingale said, "We can let circumstances rule us, or we can take charge and rule our lives from within." No truer but less realized words ever have been spoken. Remember, everything about quality of life is ultimately based on what occurs within. Outward changes and improvements can bring a momentary sense of happiness. But true abiding contentment, the ability to live well and prosper, comes from the beliefs of heart!

Having spent our lives basing everything on the external world, all the while ignoring our inner senses, makes such a statement ludicrous when faced with overwhelming life circumstances. But remember, I am not saying that the things we do on the physical plane have no value. They do! They have great value. But they are temporary solutions to lifestyle problems. Neither am I saying we should ignore the external factors. I am simply saying we ought to bring them in line with holistic priority. We can be attuned to the physical world, but let's be more attuned to our hearts.

Entering into the miraculous requires living on a new plane. It's a whole new way of approaching life. We leave the victim's world behind. We no longer see ourselves as powerlessly subject to the random happenings in our surroundings. Instead, we see ourselves as molding and shaping our world to align with our goals and intentions. We experience the reality we choose, thereby programming and energizing our inner world and our outer world to fulfill our intentions!

As a child of the Creator, living in an interactive world, your possibilities are unlimited! Having authority over the natural factors, you can learn to guide your life based on the supernatural factors. By influencing your heart you can develop faith, a deep persuasion, in the limitless power that is inherently yours. You can learn to see, feel and know what cannot be seen by the natural eye. You can sustain yourself through any situation with joy and peace, until your unseen reality takes shape on this physical plane!

Stepping into this new dimension of life and faith is not, however, involuntary! Just because you know these things doesn't mean you are immediately and effortlessly swept into this lifestyle. The great paradox is this: Creating a miracle is effortless; developing your heart to create miracles through your beliefs requires effort. It is called *laboring* to enter into rest!

> *Creating miracles is effortless; developing your heart to create miracles requires effort!*

Today you have new knowledge, tools and resources. You have in your hands the most essential keys to enter the world of the miraculous. But you have a lifetime of thinking and believing in ways that have been programmed to depend completely on the five external senses and intellectual knowledge. You have programmed automated responses that occur subconsciously when confronted with the most subtle associations. By writing new information on your heart, you can, however, write a new reality for yourself. It's like reprogramming your computer to get new responses.

Based on our sense of self, we all have programmed ourselves in varying degrees with fear, condemnation and doubt. Fear is connected to an abiding sense of unworthiness. Fear can be overcome only with love. Condemnation is the expectation of punishment, or in other words, of things going wrong. The fires of condemnation are quenched with hope. Doubt is when we believe in the things that are contrary to our hope and expectation of good things. Doubt must be superseded by faith, a deep persuasion in the thing hoped for!

People seeking an individual miracle seek only to alter their beliefs for one particular event. But the people seeking the miraculous seek to change their hearts, their self-perceptions, their life paradigms and their core beliefs so that seeing and trusting the unseen becomes a way of life. Lifestyle development is by far the most challenging initially! But it produces a lifetime of benefits. Once there is a heart change, hundreds of areas of our lives change. On the other hand, when we believe for an individual miracle without changing our hearts, we find ourselves repeatedly facing and creating the same challenges.

Change Your Heart to Change Your Life

Once while conducting a Heart Physics® seminar, I saw the perfect example of this ineffective cycle. I had led the group in a heart exercise designed to help the participants identify the beliefs that were causing them to repeat the same circumstances over and over. Afterwards we had dozens of people share what they discovered. One lady had an incredible testimonial about her beliefs that had caused repeated sickness.

She was the youngest child until her sister came along. So before that, she was the baby. As such she received all the special attention. And as would be expected, all that changed with the arrival of her baby sister. When her mother went to the hospital to deliver her soon-to-be sibling, she was shuffled off to a relative. In those days, family did not go into the delivery room and share the birth experience. This, of course, made her feel left out and a little rejected.

When they brought her mother home with the new baby, she was no longer the center of attention. Like any child, she did not know what that meant. It's not self-centeredness when others change how they relate to you with no conditioning, discussion or time for processing. So like anyone, especially a child, all she experienced was neglect and rejection. She was given no opportunity to process the reasoning!

Somewhere during that time she got sick. It may or may not have been psychosomatic. It could have been a typical childhood sickness. It really doesn't matter what the cause was. The sickness brought about a change in the behavior of the family. Now that she was sick, the family gathered around her. As any child would do, she passed a judgment. A judgment is when we assume to know *why* someone did something.[1] It is an attempt to know someone else's motive! In the mind of a little child, she believed they were showing her attention and giving her love *only* because she was sick. By default she developed the belief that people would love her only when she was sick.

A simple judgment potentially can create a lifetime of problems. In fact, most of our life paradigms were written on our hearts before we were five years old. Through the fears of a child we passed judgments that became the filters through which all the rest of our lives' experiences were

---

1 Dr. James B. Richards, *How to Stop the Pain* (New Kensington, Pennsylvania: Whitaker House, 2001), 21.

interpreted. As a child we experienced the miraculous capacity of our hearts to program our inner world, thereby charting the course for our lives! Unfortunately, until now, many did not know that we could write the beliefs of an adult on our hearts and release the destructive, inaccurate beliefs of a child.

Nothing has the power to affect us in any lasting way until we attach significance. That significance is usually based on a judgment—the assumption of why a particular thing happened. The significance we give it determines how it affects us. Because as a child she believed she was only worthy of being loved when she was sick, she set a lifelong course for sickness and disease.

Absolutely nothing so dramatically affects our lives like the beliefs we have about what makes us lovable. We will do what we believe makes us lovable no matter how destructive that action is. Love is the deepest need of a human being. Our sense of self-worth, which ultimately directs every decision and interprets every event, is the product of whether or not we *feel* loved. Love is the deepest need and the core motivator of all beliefs, thoughts and actions. So we better get this one right or there is no end to the pain!

> *Love is the deepest need and the core motivator of all beliefs, thoughts and actions.*

As this lady looked back over her life, she saw a definite pattern. Every time she felt rejected or felt the need to be loved, she would become ill. None of this occurred from conscious thought. In fact, her conscious thought was clear about the dread of sickness. Her conscious mind was acutely aware of all the pain and suffering she had endured. But hidden deep in her heart, in her sense of self, the culprit lay in wait to take her blood: a belief!

Through the course of her still young life, she had her appendix, her gall bladder, part of her intestines, her tonsils and many other body parts removed. She had suffered through many, many illnesses! The same doctors who would deny the validity of the mind-body connection were removing organs on a regular basis.

That night she discovered, in her own heart, the belief that had held her captive since she was a child. It had ravaged her body. It had cost her relationships and hundreds of thousands of dollars. So now what would she do? Would she spend her strength attempting to get a miracle for her current condition? Would she attack the problem? No! The solutions to life problems are never obtained by what we attack. She found herself in this situation because of her faith; her deep belief that if she was sick she would be loved. She had accepted a belief; all she needed to do was simply let go of that belief and replace it with a new belief!

Cleaning Out the Heart

Upon reaching this point of discovery, we go through a simple process I call *releasing*.[2] If beliefs program our cells, we don't need to attack the diseased cells. We need to change the programming. However, before we write new things on our hearts, we must erase the old.

When we release, we simply identify the destructive belief. We acknowledge that we chose this belief at some point in time. We acknowledge that we do not want this belief, that it doesn't bring benefit to our lives. And we choose to let it go! Then we go through a mental

---

[2] Hale Dwoskin's *The Sedona Method* is a great source for detailed instructions in letting go. Much of my approach was expanded by input from this resource.

image of this belief or action leaving us. Releasing is when we *put off* a destructive belief or behavior. But that is only half of this essential life-changing process.

The second part of this process is called *putting on*. Putting on is when we choose the belief or behavior with which we will replace the destructive belief or behavior! To the degree that we can deeply persuade ourselves of the value of the new belief or behavior, we will be empowered to live it. In order to create the *feelings of belief*, we must see or imagine ourselves experiencing all the possible pleasures and fulfillment of the new belief. The degree of pleasure we attach to this new belief determines the degree of our beliefs, or in other words, the effectiveness of our faith.

At first your heart will resist putting on a new belief. After all, you have invested a lifetime in a behavior that you are sure will make you happy. You have incredible faith in the belief that drives this behavior. There is something you believe you have to have to be happy and complete. You're not going to give it up easily. That is the laboring part. The only thing you can do is attach an incredible amount of pain to the belief or behavior. Anthony Robbins' Personal Power program introduced people all over the world to this process. First, make a list of all the pain in your life that has come from your attempts to get what you want. Then add to that list all the possible pain that could occur in the future if you continue. Spend time reliving the pain and shame, and with every memory remind yourself, "This pain came because I believed and did such-and-such to find happiness!" Until you attach pain to its proper source you will not even attempt to write new realities on your heart!

This concept is at the center of every inner struggle. We do the destructive things we do because of the anticipated pleasure! It's what some people call secondary gain. The Bible indicates that man was designed to live in pleasure with little inherent capacity for pain. If man was placed in the Garden of Eden, as the scripture says, we see that God created an environment with no pain, suffering or harsh temperatures. If the environment was free of pain and suffering, then it stands to reason that man would have functioned optimally in a pain-free world! Therefore, we are drawn to pleasure and we instinctively avoid pain. However, because the desire for pleasure is so dominant, we will go through pain if, first, the degree of pleasure is worth the pain; second, the perception of pain is less than the expectation of pleasure; and third, we have fantasized an incredible ending.

Perceived pleasure comes from fantasizing. Fantasizing is a form of meditation. It occurs when we imagine the pleasure we would have if we did a particular thing, married a particular person or had a particular job. Pornography is a multi-billion dollar per year industry that capitalizes on the misuse of the power of meditation. This power of meditation is so strong it is considered to be an addiction. The imagined, perceived pleasure becomes so real that the abuser faces the loss of family, income and reputation. Yet, all this repeated loss is based on perceived, imagined pleasure.

People make purchases they can't afford and face the pain of debt because they fantasize the pleasure they will derive from their purchase. Every day people make insane marriage decisions because they fantasize how happy they will be with a particular person. Sometimes it's because they fantasize the pain of living alone. Regardless of the cost, in spite of the pain, contradicting all wisdom and logic, people repeatedly run headlong into the same mistakes, the same suffering and the same failures if they are motivated toward the situation by their expectation of pleasure or if they are avoiding a situation due to their anticipation of pain!

Suffering the consequences of our beliefs and actions rarely are enough to steer us in a new direction. We always manage to displace our pain. We convince ourselves that the pain is from some alternate source. We never admit the true source of our pain! We also fail to accept one irreversible reality: *We always will repeat our circumstances if we do not change our beliefs!*

This is where the law of the seed comes into play. All seed can bear only "after its own kind."[3] Seeds are beliefs. Until you have uprooted the old seed and planted new seed, thereby developing new beliefs at a heart level, you *always* will repeat your past cycles. No matter what is different about the circumstance, what is driving you is the same, and it can't produce anything else.

> *We always will repeat our circumstances if we do not change our beliefs!*

Get in touch with the beliefs that drive your reoccurring situations. Do not make another major life decision or repeat another painful occurrence without programming your heart for a new outcome. Life is too fragile, and you have no inherent capacity to live through repeated pain. To continue without releasing your destructive heart beliefs is to risk giving up. And giving up can be anything from suicide to settling for a life of unhappiness and no fulfillment.

Release those old destructive beliefs and put on new beliefs. Spend time every day releasing fear, doubt and condemnation. Then simply put on the new, life-affirming beliefs you choose. Every life-directing belief was accepted; it can be released just as easily. Notice what's going on in your heart. At the first indication that something is not as it should be, go to your heart and identify what is driving your behavior. Make putting off and putting on a way of life maintenance.

Just as surely as fantasy can addict a person to sex, gambling, pornography or any other destructive habit, it can equally addict you to

---

3   See Genesis 1:11.

life, health, happiness, peace and prosperity! Fantasize the *pain* of the destructive beliefs and behaviors that have robbed you, and your heart will gladly let them go. Then fantasize the joys of being in the desired state until you are overcome by emotion. Then you will find yourself addicted to the best parts of life!

Now you have the power to ensure that you will never repeat another mistake or relive a past pain. You can ensure that your future is not a replay of your past!

Chapter 18

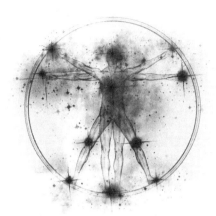

Shaping Your Brain for the Miraculous

It is written that we are "wonderfully made"![1] No words could be more true! I am so incredibly thankful that in the loving kindness of our benevolent Creator we were designed to live in the miraculous. Even though this ability may have been lost by previous generations, it is never lost from our natural design. We can delve into our hearts at any time and hear the echoes of the Creator calling us to live life on a much higher plane! It is part of who we are!

Now that we understand miracles to be laws that have been in place since the beginning of our existence, they can feel more natural. Recognizing that the miraculous does not require a commitment to any particular religious practice eliminates the dread of a life bound to ascetic religious idealism. Understanding that miracles are not the violation of natural laws, but rather natural laws we do not yet comprehend, places them well within the reach of possibility. Realizing there is no conflict between science and faith, just misunderstandings, frees us to examine both realms with no fear that one will corrupt the other. Because modern science has given us a language that makes sense in our generation, we

---

[1] See Psalm 139:14.

can explore the miraculous without having to turn off our brains! Instead, we can investigate the miraculous and develop our brains for a greater capacity to live in the realm of the miraculous. Choosing to live in the miraculous is the most freeing and liberating choice we can ever embrace! It is the opportunity to live how we were created to live!

Shorter Exercises Bring Longer Benefits

We all have heard the saying, "Use it or lose it!" Well, it's more than just a saying. Muscles we do not use will weaken, decrease in size and eventually atrophy. This concept also holds true on our other levels of existence. In the efficiency of our design, that which is not used is deemed as unimportant and perhaps even disposable. Conversely, that which we use grows, expands and is strengthened. It is strengthened specifically the way we train it. For example, new discoveries indicate that the way we have been taught to exercise trains our bodies to keep replacing the fat we lose. That's why so many people begin to gain weight immediately after they stop exercising. Amazingly, there are breakthroughs that show that shorter exercise time, utilizing slightly different techniques, train our bodies to efficiently burn fat, increase our metabolism and keep the fat off! Everything we do repeatedly programs our brains and our bodies to function a certain way.

The same is true of strengthening the heart. It could be that the aerobic craze of the past four decades has played a role in the rise of heart attacks. Aerobics trains and shapes the heart for a way of functioning that we never use in daily life. How often do you have to jump up and run five miles in daily life? Never! However, there is much research that presents an almost unarguable case that short bursts of intense effort for a very short duration causes the heart and lungs to actually grow in a way that is more useful in daily life, thereby reducing heart attacks.[2] The end result

---

[2] Al Sears, M.D., *PACE: Rediscover Your Native Fitness* (Wellington, Florida: Wellness Research & Consulting, Inc., 2007).

is that you train you body for the kind of strength you need in daily life: quick short bursts.

The way we habitually think teaches us to become more aware of pain. The way we repeatedly react to certain types of situations trains our bodies to put us into the fight or flight mode at the first sign of such a situation, thereby increasing the amount of certain hormones to very dangerous levels. The way we think can program us to eat at the wrong times, store fat instead of generate energy, stay awake all night and be sleepy all day. We train our brains to release chemicals excessively, which lead to chemical imbalances. We train our bodies to crave sweets. We program ourselves to break every health regimen within a predictable time. We even train our brains to make us depressed or happy! Everything we do repeatedly, good or bad, becomes a process of programming.

> *Everything we do repeatedly, good or bad, becomes a process of programming.*

The same is true of our inner senses. Few of us have developed our inner senses to any functional degree. At the end of the day, our entire lives are guided by our intellect, with little regard to the unending stream of intuitive signals that come from our hearts! Once in a blue moon we recognize the deeper knowledge. We suddenly, without thought, stop at a green traffic light for no known reason, just in time to avoid a deadly collision with a speeding vehicle that ran its red light! We don't know how we knew to stop. *It's a miracle!* Or so we think! But because of our definition of a miracle, we do not recognize the working of our inner senses as an ability we could tap into at will.

Highly developed martial artists, on the other hand, can feel the energy of a person with the intention to strike them. Some people can *feel* impending danger. Others *sense* when they are not hearing truth. Through hypnotism, surgery patients control bleeding while undergoing dangerous operations. In meditation, highly developed people can lower blood pressure and slow respiration to an unperceivable rate. I have seen highly developed martial artists lay all of their body weight onto the tip of sword and never break their skin. All of these are simply skills, latent within every human being, that have been acquired through the development of the inner senses. As we discovered previously, it is where we place our attention and on what we place our attention that map and program our very beings.

Our Brains Grow

In the book *Evolve Your Brain*, Joe Dispenza cites a study done at the University of Regenburg in Germany.[3] Subjects were taught to juggle. In three months the subjects who learned to juggle had measurable growth in gray matter in the parts of the brain involved in visual and motor activity. They didn't simply acquire a new skill; their brains grew to give them new capacities for that skill.

When we meditate and see ourselves as well, prosperous, active, energetic, happy or whatever we choose, we set off an avalanche of miraculous events. Our brains grow to give us new capabilities to do what we see. Our brains create neuro-pathways, release hormones and initiate possibly millions of other processes to make this miraculous ending a physical reality!

---

3   Joe Dispenza, *Evolve Your Brain: The Science of Changing Your Mind* (Deerfield Beach, Florida: Health Communications Inc., 2007), 57-58.

Our brains grow to give us new capabilities to do what we see.

Remember, the mind knows no difference between what we clearly imagine and what we experience in the physical world. After all, whether real or imagined, the experience is the same. The brain releases the same hormones and chemicals for the imagined and the actual. Any emotion we experience is real to our minds! That's why when we *feel* afraid, whether real or imagined, our bodies shift into the fight or flight mode and thousands of bodily processes change. Doing this regularly, even though it's not real, can totally destroy our health by an imbalance of chemicals that prepare our bodies to meet an imaginary threat!

It is common knowledge in many circles of the experiments that were done where one group of people shot basketballs from the free throw line, while another group imagined shooting from the free throw line. Both groups did their *exercise* for the same duration. At the end of the study, when the group who only imagined came onto the court, they had improved almost as much as the people who actually practiced. Their brains created the capacity for the skills as if they actually had practiced every day!

Turn your brain and your body into an organism that is alive to the miraculous. Spend time *every day* being who you want to be, how you want to be, feeling the way you want to feel. See the end of your journey from the beginning! See, imagine or visualize yourself living the end now. As you do, you will program the cells of your body to conform to that image. You will charge your outer world with the influence to become what it needs to become. You will attract to you all that you need to fulfill the dream. Your thinking will change to correspond to that person you desire to be. And so, seemingly unimaginably, you will grow and develop the parts of your brain that are needed to function as that person with that life!

Your miracle is not simply mind over matter. It is mind creating matter—gray matter, muscles, neuro-pathways, cells and billions of other components—making it possible for you to *be* the person you *see*! The ultimate miracle doesn't simply give you what you wish to have; it makes you to be the person who can have that life every day!

Chapter 19

The Law of Harmony

We all will choose to live one of two ways: from our hearts and intuition or from our five external senses and our intellect. Most will and should utilize some combination of both. We are, after all, in a physical world. That physical world is, however, ruled by the invisible world. So we too will be dominated by one or the other. When it gets down to making decisions, where does the buck stop? Are our hearts or our heads the final factor?

The word *conscience* is one of those words that we loosely throw around with little real understanding of its meaning and value. We tend to think of the conscience as the feelings of guilt we experience when we do something "bad." In some nebulous sense that could be true. The greater meaning of conscience, however, will serve to usher us into a life of harmony!

Conscience is a compound word that basically means dual knowledge. Through our five senses we have knowledge of the physical world. Through our inner man we have knowledge of the invisible world. Just as the physical body constantly gathers data based on tangible input, so the inner senses gather *energetic* input. The input of the five physical senses is evaluated by the intellect. The energetic input is known through the intuitive knowledge of the heart.

When there is disagreement in the conclusion reached by the evaluation of both sets of data, we are conflicted in our conscience—our dual-knowledge. Through the lack of harmony in these two realms we feel *dis-ease,* which leads to stress, guilt or other negative feelings that physically manifest as *disease.*

The (In)ability of Force

The final source of your evaluation determines if you will live by the laws that dominate the physical plane or the laws that guide the invisible plane. The way of the physical world is force; the way of the invisible is power! The two seem very much alike to the casual observer, but when looking at the anatomy of a miracle, you find all the difference in the world.

Those who live by the inner, miraculous laws of life easily can harmonize the internal with the external. It is after all the internal, invisible world that rules the external, physical world. Those who are ruled by the external cannot, however, harmonize those laws with the internal laws of miraculous life!

In the physical world we use force to overcome the natural laws. There is no respect or understanding for the unseen. We attempt to reach our goals by overcoming the natural laws. Ignorance, pride and impatience see no value in applying the laws of the miraculous. We are so *natural* minded that we find ourselves opposing the very supernatural laws that were given for our good!

When we violate the natural laws, there is no end to putting forth effort. We have to generate enough force to overcome inertia. Then, even though it may get a little easier, we have to keep using force to keep up the momentum. Regardless of what we do, in the end when we stop putting forth enough force, it all comes crashing down. There is no rest or peace in a life of force; only conflict, fatigue and stress.

> *In the end when we stop putting forth enough force, it all comes crashing down.*

Airplanes are a great example of a force that overcomes the laws of nature. By creating enough thrust, combined with lift, we amazingly can get several tons of aircraft and passengers into the air. Once it reaches cruising altitude, it takes much less force to keep it moving. But be assured, when there is no more thrust, it will come crashing to the ground! In the same way, a life that violates the laws of the miraculous eventually crashes under the weight of its own success!

Yin and Yang

The way of the miraculous is not to violate the laws of nature but to harmonize with them. By harmonizing with the supernatural laws of life we harmonize our conscience, our intellect and our intuition. We in the West have little concept of the law of harmony. We tend to do everything by force. It is this very disharmony, however, that creates much of our need for the miraculous as we now know it (which is a supernatural deliverance from the problems *we* created through the use of force).

Chinese medicine is rooted in the concept of yin and yang, or polar opposites. Just as planet Earth needs the rain (yin) to nurture the plants, it also needs the sun (yang) to cause them to grow. Just as the earth needs the harvest (yang) to make room for future abundance, so it needs the winter (yin) for the soil to rest and be nourished.

In the Old Testament, the Hebrews were taught that honoring this principle of rest (yin) and harvest (yang) was essential to abundant crops. They were taught not only to allow the land to rest during the winter, but also to abstain from planting anything every seven years. Harmonizing with nature was one of their keys to prosperity through abundant crops! Additionally, while the land was resting, they would be able to rest and recover their own health!

Observing the principles of yin and yang, when we need more energy (yang), we should rest (yin). Instead, in the Western twenty-first century lifestyle, we drink caffeine (yang), take guaranna (yang), or ingest some other stimulant (yang). Since we don't get energy from our much needed restful nourishment (yin), we have chronic fatigue, heart disease and a host of other diseases that come from pushing our bodies into extreme acidic inflammation (yang). Almost all modern diseases are the product of stress (yang), which causes acidic inflammation (more yang), which facilitates nearly every form of disease (extreme yang). Resting (yin) one day out of seven, as prescribed in the Old Testament, was a practical application of this very principle. A life in harmony balances the yang and yin foods, activities and energies for perfect health.

If we refuse to harmonize, we exhaust yang energy. When we can't push ourselves any further, we collapse into extreme yin, the total inability to function. When we don't willfully harmonize with the miraculous laws of abundant living, we experience adverse consequences. It is not a judgment from an angry God. It is not the universe getting even. It is what happens when our force runs out. Our airplane simply comes crashing to the ground! An innumerable host of modern diseases simply would disappear if we only took more vacations, relaxed more often and learned how to play.

> *Many diseases simply would disappear if we only took more vacations, relaxed more often and learned how to play.*

Our lifestyles, produced by our mindsets, put us in disharmony with the laws of nature, thus giving rise to our countless woes. The refusal to give up our mindsets keep us from entering the miraculous and experiencing a miracle. The Western ego-driven lifestyle, which is exported to all the world as the ideal way to live, is completely toxic! It is a lifestyle that works longer hours with fewer vacations and off time than any other culture in the world. What we actually are exporting to the world is a legalistic, puritanical work ethic driven by greed and laced with life-threatening, stress-related diseases that fails to honor the God it claims to believe in. Holding to these unhealthy ideals keeps us in a cycle of sickness and distress!

The refusal to change our lives when we somehow break through and receive a miracle keeps us from living in the miraculous. When living in the miraculous we do not continue repeating mistakes, creating problems and seeking deliverance through miracles. That is the way of the foolish! No! We harmonize our lifestyles with the laws of the miraculous and live in an entirely different quality of life.

This is where some struggle with the accurate concept of "a miracle." We don't consider it to be a miracle when a person tends to make things work and stays healthy, prosperous and happy. We just consider that person to be lucky. But would we not consider it to be miraculous if we had spent our entire lives sick and, by operating a few principles, we got well and stayed well? Would it not be a miracle for a person who struggles a lifetime with financial difficulties, to harmonize with the laws of prosperity and rise from his sad state? How much more miraculous is the person who avoids these calamities by trusting and harmonizing with the laws of miraculous living!

Harmonizing requires entering into the *state*. Enter the state of peace and feel your body change. In time, living in peace will nurture your physical body and you'll have more energy and stamina. Enter the state of prosperity and notice that you no longer feel desperate. You start making better business decisions. Enter the state of health, and you'll find yourself saying no to junk food. You'll want to walk and move more.

Whatever *state* you enter, by harmonizing your thoughts and actions you will not only feel an internal difference, but you also will notice that the things you need to live in that state will find you. The people, connections and resources you need will be attracted to you. And just as it takes no effort for a magnet to attract the iron fillings, so you will effortlessly attract the natural complement to your inner thoughts and beliefs.

When you harmonize you thoughts, beliefs and actions with the immutable laws of the miraculous, you will live a miraculous life! As you honor the laws of harmony, you will recognize that you are pulling out of the old life of pain and suffering. In time you will notice that you win more than you lose. As time goes by your life will get better and better. Less of your energy will be devoted to solving problems and more of your energy will be devoted to new life, new realities and new joys!

Chapter 20

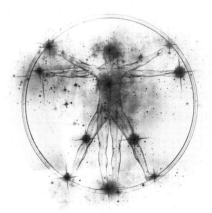

Harmonizing Your Outer World

Just as we must harmonize our inner worlds to arrive at a predictable outcome, so we must harmonize our outer worlds as well. There must be congruence between the end we desire, our thoughts and feelings, and our words and actions on the outside! The greater the disharmony is, the more arduous and prolonged the process becomes.

Belief and action are a continuum. In some of the earliest writings about the power of believing, the words *believe* and *obey* are synonymous. We do, in fact, obey our beliefs in the form of corresponding actions. People who say they believe for a particular outcome yet whose behavior is incongruent probably have an intellectual acknowledgment, not a heart belief. Or, they have a stronger opposing belief (doubt) that is guiding their actions! Make no mistake, however, our actions always reflect our strongest *felt belief!*

Even in basic goal setting, the leaders of the motivational field insist that you put corresponding actions with your goals. Listen to the words of Anthony Robbins, internationally known motivational speaker and life coach:

The most important thing you can do to achieve your goals is to make sure that as soon as you set them, you immediately begin to create momentum. The most important rules that I ever adopted to help me in achieving my goals were those I learned from a very successful man who taught me to first write down the goal, and then to never leave the site of setting a goal without first taking some form of positive action toward its attainment.[1]

In beliefs of the heart, actions and outcome cannot be separated from our sense of self, from our personal identity! Therefore it is necessary that we begin to live, act, talk, think and even dress like the person who lives the end result. It is unavoidable! In the continuum of beliefs and actions, one reinforces and empowers the other. Our actions are signals to our hearts that our beliefs are real. They become an essential part of the ability to receive!

Live What You Believe—Now!

One of the writers of the New Testament said that faith without works (corresponding action) is dead.[2] In other words, faith, which is being fully persuaded, is completely useless apart from *belief*, feelings and thoughts, which compel us to congruent actions! When we realize that most of the miracles we need are found in a state of being, it becomes completely understandable that our actions would become harmonious with our expected miracle! A person can do without believing, but a person cannot believe without doing! Living in a state brings about congruence in every aspect of our beings!

---

[1] Quotes by Anthony Robbins <<http://www.achieving-life-abundance.com/quotes-by-anthony-robbins.html>> December 9, 2008.
[2] See James 2:20.

Beliefs of the heart are inseparable from identity. In other words, we're not just getting a miracle. We are *living* the miracle. We've got to be that person. For example, if you can't *be* the man Mrs. Right wants to live with, you'll mess it up if you catch her. Being is more than an event; it is a lifestyle. Being is about more than getting; it's about having! The state of being is not a goal one reaches by doing. It is a state wherein one lives by being! The man you are can't live with Mrs. Right; if you could, you would be! An ancient proverb says, "The prosperity of fools shall destroy them!"[3] To obtain without being will ensure the loss of that which is obtained!

If an insecure man catches the woman of his dreams without becoming *that man*, then jealousy, insecurity and control will be employed to hold onto who he feels unworthy to have. If you are not Mr. Right, when you meet Mrs. Right you won't be the man she wants. People mask their identities when dating. Guys pretend to be the man Mrs. Right wants. But that level of energy cannot be maintained. In time who he is cannot be contained and the end of the relationship begins!

You have to be the person who intends to live in the miracle!

If a man who feels like a pauper suddenly wins the lottery, he will either, squander it all, make ridiculous investments or live in greed and fear of losing the windfall he doesn't feel he deserves. You have to be the person who intends to live in the miracle! When chance, manipulation or sheer determination delivers into our care a life beyond our true self, its upkeep will be our downfall unless we develop ourselves!

---
3   Proverbs 1:32 KJV.

The language of the heart is always "personal and present tense." Believing means, "I am experiencing the end now; I have what I desire in non-tangible form now. If I have the desired end now, if I have entered into the state I desire, it would be a denial of my belief to live as if I do not!" Wallace Wattles says it like this: "By thought the thing you want is brought to you, by acting it is received."[4] We will begin to make preparation for anything we truly believe is going to happen. The person of faith makes preparation for what is already his, in his heart experience!

Being Prosperous

The person who is *getting* the miracle of prosperity at some time in the distant future, doesn't live, think, feel and make decisions now like a person of prosperity. In fact, such a person makes decisions that may send the miracle away when it comes! If someone attempts to transact a business deal with the person who is the other half of the miracle and that someone isn't living in that reality, he may well compromise the transaction. The person who is fully persuaded feels and lives like it is his now, and he interjects that into every relationship and transaction.

So many miracles involve meaningful interaction with others, possibly none more than prosperity. Money is not going to float down from the air to make us prosperous. It is unlikely that it will be given to us, unless it is for a cause that will serve others. Since prosperity is a state of being and thinking, to have money given to us would not make us prosperous. It would create a welfare mindset! It could defeat the entire process if money were simply given to us.

A prosperous person is a person who sees opportunity in any situation and takes the definitive steps to reach the goals. The truly prosperous is never competitive with others. They know there is more than enough to go

---

[4] Wallace D. Wattles, *The Science of Success, The Secret to Getting What You Want* (New York, Sterling Publishing Co., Inc., 2007), 69.

around. The prosperous rely on creativity, not competition. The prosperous at heart feel that wealth is good for everyone and that there is always enough for everyone to have. They are as committed to the success of others as to their own success. They have the sense of *more than enough*... more than enough money to go around, more than enough opportunity and more than enough no matter what the economy is like. And in the event prosperous people suffer loss, they know there is always enough opportunity to earn it back! They don't waste away bemoaning their losses. They rise up from the power of their prosperous hearts and earn it again!

Prosperity is about *more than enough*! When wealth (a goal) is confused with prosperity (a state) the idea is that we always have to have more than enough money. No one ever has more than enough money unless they have prosperous hearts. A truly prosperous person is able to so enjoy what he has that he can be truly happy in almost any situation. Ultimately, he knows the situation will conform to his state!

Many people do not know how to recognize the miraculous when it occurs. It is something they have never lived. It is completely foreign to them. Become fully associated with living in your miracle. Experience it every day in your heart as you meditate. Make it real now! Then live in this world as you will live in your miracle. Walk the walk! Talk the talk! Be that person now, as much as it is possible.

Be the person living the miracle NOW.

Receiving, then, is not passive. It is *taking hold* of the person who can live the miracle, in our hearts! Some pray and wait in idleness. Some think that because they want it bad enough it will happen. Some think that just thinking about it will attract it. Although all of these ideas embody a partial truth, Jesus, the Great Miracle Worker, always called people to act on their faith.

Everything we do or don't do plays a role in persuading our hearts. If our words deny our hopes, then our hopes will never become beliefs. If our actions deny our beliefs, then our actions will create doubt. If our doubts confirm our fears, then the thing we fear will become our reality!

Besides the incredible influence that occurs in our own hearts from the congruence of our actions and our beliefs, there is the additional bonus of the outside world. As other people see us living in the reality of our miracles, they come to our aid and support. When we live in the defeat of our situations, we draw people who enable more defeat. They give us sympathy instead of support. They even prey upon our manifested weaknesses. But as we live in our miracles, people come alongside us to help us live the life they perceive us to have chosen. There is no limit to the things that begin to go our way when we harmonize our inner and outer worlds with the end we desire!

But here is the ultimate miracle of being. We don't really know how to *be* the person we have never *been*! We could spend years studying the lifestyle of the person we hope to become. But that knowledge would be little more than an attempt at intellectual imitation. Information alone does not contain the power of transformation! But we don't have to know all the facts. If we can see ourselves *being* that person, our hearts will develop the beliefs, our minds will think the thoughts, our brains will grow the pathways and the character of the person we desire to be will emerge from our hearts! Transformation starts from the inner world of belief. And belief will guide our actions to the life we have chosen!

Chapter 21

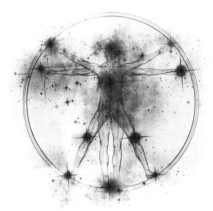

The Language of the Heart

The heart communicates in very specific ways. Realizing and harmonizing with the way information is taken into the heart and the way it is released from the heart is key to a positive, proactive, deliberate quality of life! Even many people who have an accurate grasp on the role of the heart may not understand how to best utilize this information for optimal results! But the initial rules are simple.

As we already have discussed, words are so very powerful. They are the seeds that are sown simultaneously on all levels of our existence. Words, when harmonized with belief, feeling, and thought, may possibly influence more planes of reality than any other thing we do! The *mouth tends to speak what is really in the heart*, even when we don't want to do so! Although we may temporarily mask what is in our hearts, it usually comes out, with feeling!

An ancient proverb tells us that life and death are in the power of the tongue.[1] Our lives will be filled with and influenced by our words more than any other thing. As previously considered, words not only influence

---

[1] See Proverbs 18:21.

ourselves; they also influence others. There is, therefore, a multiplication that occurs with our words, on so many realms. That may be why the Proverbs of Solomon, considered to be the wisest man alive of his time, contains more warning about the use of words than anything else!

Words are seeds sown on all levels of our existence.

We know that words affect us internally while simultaneously affecting the people around us. The primary difference between words that are spoken and words that are thought is that our inner dialogue doesn't influence others. It still powerfully influences us! As people of free will, we constantly are choosing our direction and quality of life. That choice is reflected in our self-talk, especially the self-talk that creates emotions. Remember, information (self-talk) at times of strong emotion results in writing on the heart! Having come this far in the book, we have little need to discuss the effects of default thinking. Instead, we should now focus on some of the known ways to positively influence our hearts.

The first and most fundamental rule for writing on our hearts is that all statements, whether verbal or nonverbal, must be personal, positive and present tense! The heart is not rational. It simply gathers and programs information the way we say it or interpret it. As such, how we say it is as important as what we say—sometimes even more important! Violating these rules of self-talk not only will fail to garner the positive results we desire, but also possibly bring destructive consequences.

Make It Personal

The first law of heart language is this: It must be personal! Everything you choose to believe at a heart level must be in the first person. At first glance this seems rather easy. Never say "we" or "us" or be general when using self-talk. Never generalize inner statements. Every sentence spoken to influence your heart must begin with "I," and "I" must be at the center of every picture or visualization.

When you include others in your faith statements, you actually create boundaries and limitations. Let's say you include another person in your picture of happiness. You easily could attach so much significance to that person being there and conducting himself in a certain manner that you do not believe you can be happy without that person.

In advanced heart work, I teach people how to include someone without making that person's presence become a boundary! But few people know how to do that. Remember, so much of what improves your quality of life is a state, not a goal. Therefore, no one else has control over your *state*. That is completely internal; therefore, it can involve no one outside of you. In your heart you must be able to experience the end independent of any person or specific event!

Several subtle but very destructive things occur when you include anyone or anything in your picture. As already mentioned, you believe you must have that person or item to reach the desired state. It becomes a barrier to your desired end result! If you think you must have someone to be happy or successful, you will focus on his behavior. You can slip easily into control or manipulation. You become a taker in the relationship. You need that person to have what you want. You have placed your focus on something over which you have no control: another person. Always deal with the one thing you do have the authority (the right) to deal with: your own thoughts, beliefs and actions.

If you want you and your spouse to be happy, you only have the capacity in your heart to deal with your part of the process. See yourself being a loving, positive person. See yourself handling conflict in a healthy, positive way. But don't see yourself happy because your spouse responds to you or treats you in a specific manner. In other words, see yourself being happy and you can develop the state of happiness independent of your mate's choices, health or presence in your life! Then you bring happiness to the relationship. Instead of force, you now have the power of influence!

Stay on the Positive Side

The second aspect of heart language is this: It must be positive! What we start is far more important than what we stop. For example, consider a self-statement like this: "I only eat foods that contribute to my health, energy and longevity." Without addressing anything negative, you have made this reality personal, positive and present tense! Every part of your being will be motivated to fulfill this statement.

You could take the next step and put it with an identity statement: "I am healthy and energetic. I value my health and energy." And then you could attach purpose to it. "My health and energy give me the capacity for health, wealth and a wonderful sex life. Therefore I only eat foods that contribute to my health, energy and longevity." A statement like this coupled with a picture of you being healthy and energetic, coupled with strong emotion…your battle with junk food is over!

Because of our *nature* to pursue pleasure and avoid pain, our hearts will not accept or be motivated by or toward negatives that would be viewed as pain. But remember, if we attach great amounts of pleasure to something destructive, or if we state something in a way that emphasizes the secondary gain, our hearts may accept it as a desirable!

While studying psychology, I once read a story about a young girl who was very sick. So the emotional environment was charged due to her age and the emotions that presented themselves during times of sickness were very strong. Within the hearing of the child, the doctor made this statement to her mother: "She will never live through this!"

The little girl heard the statement and developed this belief: "If I ever get *through* this sickness, I will die." That's not what was said, but it was the interpretation she gave to it, and it is the way the belief was written on her heart! After years of chronic sickness, this belief was revealed in a hypnosis session. Releasing such a belief and writing something new on the heart ends these types of lifelong struggles.

Releasing negative beliefs and writing something new on the heart can end lifelong struggles.

The secondary gain of being sick (pain) was staying alive (pleasure)—or so she believed. In the same way, we understand the secondary gain of a woman who was molested as a child. During a time of extreme emotions she was told that it was an act of love. As an adult she may hate promiscuity; it may violate all of her beliefs and bring enormous shame and despair, but she conceivably could spend her entire life struggling with immorality (pain) in her search for love (pleasure). Because of the words spoken, sex and love became synonymous. The pain of immorality was trumped by the illusionary secondary gain of finding love.

A new belief about living healthy or a new belief about being lovable as a person will end the need to run the old default programs of dysfunction. As adults with life experience, we no longer need to operate from the heart beliefs established as children, influenced by those who hated us or sought our harm. Friend, for the first time you have the keys to living the rest of your life the way you choose. There is no part of your life you cannot change. There is no habit you cannot break.

Now Is All We Have

Finally, all heart language must be present tense! Now is all we have. Since the heart deals with our sense of identity, everything must be in the *now*. We have no future identity. All actions must be linked to beliefs that exist in the now. To say, "I am going to quit smoking!" Is a perfect example of going about the process almost completely wrong! It has been made personal, "I am." But "going to" is linked to some future time. "Going to" indicates something we must do instead of who we are. Since that is in the future, it is an unknown that we cannot experience now. All unknowns are considered to be pain! Likewise, effort equates pain. Quitting is negative. And since many smoke because they enjoy it, they will perceive quitting to be the pain they don't want and smoking as the pleasure they do want! This internal statement of "going to" will create a conflicted state and may very possibly make someone desire to smoke even more.

Do you want to quit smoking? See yourself healthy and strong. See yourself saying no when a cigarette is offered to you. But see yourself smiling and feeling good about your self-control. Then make positive statements about your health and longevity. Make statements about living and being healthy to spend time with your kids or your spouse. Make all of your statements personal, positive and present tense. Then see yourself living the way you want to live. Attach every possible benefit. Experience feeling the way you want to feel.

Wanting a desired end but failing to speak the language of the heart would be tantamount to explaining chess to someone who spoke another language and thinking he would understand the game. Speak the language of the heart and you will see more of your desires fulfilled. But the key is not to simply speak a proper inner dialogue while in meditation. Speak the proper heart language at all times. Make all your inner communication something that supports the desired end. And make all your verbal communication harmonious with the inner dialogue of faith. In so doing you will not waver; you will not fail. You will be apt to lose touch with the old beliefs of the heart and you will see the end come to pass!

Chapter 22

Heart Physics

Over the past thirty years I have developed a program I call Heart Physics®. This program is based on the theory of microcosm–macrocosm. This theory, in part, presents the idea that all things in the universe operate by the same governing principles from the largest to the smallest part. Thus, the laws that govern physics are the same laws that govern the heart.[1]

One of the first and most important laws of Heart Physics® comes from the idea that for every action that is an equal and opposite reaction. This is Newton's third law of motion. This law is essential for congruent belief and heart language. It is through the misuse of this law that many negative and destructive beliefs are written on the heart, even when there is intent to do just the opposite.

It is applying this law that makes us better grasp and utilize the law of personal, positive and present tense! It is here we begin to see the subtlety of our inner language. The way we make certain statements creates an opposite effect in the heart. Let's look again at our earlier example of

---

[1] For more information, go to www.heartphysics.com.

"I am going to quit smoking." "Going to" implies sometime in the future, which means it is something that the speaker does not presently have. If he does not presently have it, then it is not a reality to the heart. It is a hope or expectation, not faith (complete persuasion). Our hearts cannot give us the strength to do something that is not present tense. If we are not completely persuaded, we will waver.

When applying this law, we realize that to say we *want to get something* is to say we do not have it. I have heard people say, "I am getting healed," or "I'm going to be healed." Although they may be positive and optimistic, they fail to realize that to passionately say, "I am going to be..." is to passionately say *I am not*. All of the effort put into establishing faith can be doing just the opposite. Instead of drawing closer to the desired end, we may be pushing ourselves farther away. This is why all affirmation must be positive and present tense. That which we do not have in the present is a reminder that we do not have it!

That which we do not have in the present is a reminder that we do not have it!

Being and doing is another area where this can sometimes make a difference. Unless we attach massive amounts of pleasure to doing something, we often can dissuade ourselves from the task. Remember, *doing* often represents work or effort. It is often the dread of effort that has kept us locked into a particular problem.

In making our affirmation personal, we must emphasize being more than doing. Now here's a real nugget. If we see ourselves healthy, slim and trim, that is a state of being. The fact that it is present tense causes our minds to release all the hormones and chemicals needed to make it a reality. Remember, the mind knows no difference between what is imagined and what is real.

Then, if we allow ourselves to experience all the joys of being so incredibly healthy, it is now positive. As a result, we are experiencing the *state* of health. Unless we desire to or know how to do it and have a positive effect, we do not have to see ourselves exercising or eating a particular way to reach this state. All of that is about becoming, not being.

The Problem With Becoming

"Becoming" is one of the most devastating concepts we can embrace. First of all, the need to become implies that we are not (an equal and opposite reaction). That removes it from the present tense. But the greater destruction is the defamation to our sense of self. To suggest that we need to become is a rejection of who we are. It seems the law of the paradox applies here.[2] We must embrace who we are in order to grow into who we want to be. We must be, to become!

The heart and mind are so complex and so powerful, yet so predictable. We can forget about all the steps of *becoming*. We can eradicate the terminology of becoming from our inner vocabulary. We can see or imagine the end as a present tense reality. Our minds can relate to our bodies as if the desired end is real. Therefore, it can make every part of us function as if that is the reality.

If we are seeking to become or do what we have never been or done, it is obvious that we do not consciously know how to do it. A life of growth and abundance is a constant journey to where we have never been. Since we have never been there, we probably don't know the way. We don't

---

[2] The law of paradox states, "Reality always lies in the balance of two seemingly opposing truths." (James B. Richards).

know what our bodies will need, but we can trust our hearts to guide us. But, if we insert particular steps into our meditation, we are not seeing the end; we are viewing the process. And what's worse, it could be the wrong process.

If we see ourselves as healthy, our minds will guide the behavior and biological processes of healthy eating and exercise. If that is who we are, then that is what our minds will lead us to do. If we see ourselves as prosperous, we will make the types of decisions that a prosperous person will make. Our efforts should focus on *being* and letting our hearts figure out how to get there.

Desire, when not taken in a self-destructive direction, can be a positive thing. When we *want* to do something, it is much easier to do than when we don't want to do it. When we desire something, the labor involved in the journey is a joy. Sometimes, however, we do not distinguish between what we desire to do and what we are obligated to do. The difference is massive on an internal scale. The subtle evidence of obligation versus desire is often identified in the terminology.

Ought Means Not

For example, if we use obligatory terminology like *ought to, have to, should* or *need to*, we are making the statement to our hearts that we do not "desire to." We are implying that the action we are about to undertake is done solely on the basis of obligation. "Ought to" implies, we don't really want to! At a heart level we will resist that which is motivated by obligation. Obligation is pain! The heart avoids pain whenever possible!

When you hear these negative words in your inner dialogue, stop immediately! Make a conscious decision about why you are doing what you are doing. Is it moving you toward a goal or desire? If so, state the reason for taking the action. Immediately connect to the pleasure and fulfillment that will come to you by taking the step you are about to take. Then change your words. Restate the intention: "I *want to*" or "I *desire to*." You possibly should take a few seconds and reaffirm the positive reasons for the action.

Sometimes obligatory or negative terminology reveals our true belief. I often have had people in my seminars, during personal conversations, make statements like, "I know I *should* do Heart Physics® more consistently." I often stop them with the question, "Do you want to, or not?" "Well, of course I do," is the standard reply.

When questioned about the reason they want to do more, it's like giving a fifth grader a science test. They have memorized all the correct answers, but they have no idea what it really means. Usually further investigation reveals one of two things: Either they are not really persuaded of the benefit this will bring into their lives, or they are not having positive, enjoyable experiences during their heart work. Until we have enough positive experiences to be fully persuaded, there is no positive belief (thought and feeling). In such a case there is only information and obligation!

I realize that a person at this stage of the journey is operating on will power. Will power is a great starter, but it's not much of a finisher! When there is a struggle between will power and feelings, feelings always win! We use choice and will power at the beginning of any new endeavor. But very early on we must have positive enjoyable experiences or the will power fades away and we make a new decision.

Most people train their hearts for failure. Let's say we start eating healthier. We're only doing it because we feel obligated. We are more attached to the pain than the pleasure. After a few days or even a few weeks of pain and suffering, our will power wanes and we give up before having any positive experiences. The next time we consider eating healthy, we have to overcome the reinforced belief that it is all pain and we will not see it through. Every time we start and quit something, it gets harder to try again. In my Heart Physics® program, "Living Healthy," we teach people how to attach to the positive reasons to be healthy and to experience it immediately. This has been the secret to empowering people to lose weight, exercise and eat healthy.[3]

Attaching positive reasons to being healthy and experiencing it immediately is the secret to empowering people to lose weight.

This is where another law of physics comes into play. Newton's first law, the law of inertia, says, "A body continues to maintain its state of rest or of uniform motion unless acted upon by an external unbalanced force." To get a body moving from a resting position, or to get a body in movement to change its direction, it must be acted upon by a greater external force!

You may be "stuck" in a particular arena of life. Or you may have been entrenched in a particular problem for years. In order to get you moving or to change your course, you need to be acted upon by an influence of great force. Unless you can create a stronger belief (thought and feeling) through heart work than you are experiencing through your current situation, you will either remain dormant or stay on your current course.

At the beginning of any new endeavor we must experience hope—positive expectation of good. It is rare that a person experiences an instantaneous miracle, but not impossible. The more something is

---

[3] See the module Living Healthy at www.heartphysics.com.

totally limited to our own physical beings, the more likely we are to see an instantaneous miracle. But so many situations, like finances, require the involvement of others. They are never instantaneous. In lieu of an instantaneous miracle, we must have an incredible positive *instantaneous experience*—something that makes the end very real and believable. Faith (absolute persuasion) springs from hope (confident expectation). Evidence gives birth to faith and hope. We must create the evidence in our own hearts. And we must make it more real than the current state of being!

To experience the miraculous—a miraculous change in the way your mind and body work, a miraculous change in the world around you or a miraculous change in your condition—you have to have a belief (thought and feeling) that is greater than your belief (thought and feeling) in the current opposing reality. See and experience the chosen end from the beginning. Say internally and verbally the end you chose. Stay connected to the (feeling and thought) belief in that end. Do not doubt; that is, do not give your belief (thought and feeling) to the opposing circumstance. Then you will have what you say (verbally and nonverbally).

When you learn the keys to Heart Physics® you have all the tools you need to reprogram your world, break destructive habits and enter into an incredible quality of life!

Chapter 23

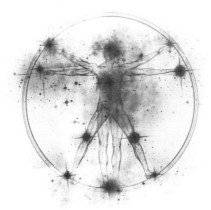

Persuading Your Heart

One of the most essential paradoxes of the journey to the miraculous is found in the ancient saying, "Labor to enter into rest."[1] Such a statement at first seems to be contradictory and even nonsensical. But like all reality, it is understood only within the paradox! Living in the miraculous is an effortless place of rest. *Getting* to the place of rest does, however, require labor.

Remember, faith is when we are fully persuaded of the things that, up until now, we were confidently expecting (hopeful) would come. This deep persuasion is the evidence that the thing we cannot see yet really exists and really is ours! Hope and faith always are intimately connected. One is essential for the other. Hope (positive expectation) is the mother of faith (complete persuasion). Faith, like any responsible child, grows to provide for, support and nurture its mother! Without faith, hope eventually wanes. Without hope, faith is never born!

Faith and hope are complementary opposites that form the paradox wherein the miraculous occurs! Through the course of the entire journey

---

[1] See Hebrews 4:11.

there is hope—a confident expectation of that which is to come. But at the same time there is faith—a deep sense of present tense ownership! It's like the positive and negative poles in the filament of a light bulb. The light dwells in neither pole. It only exists in the space between the poles. Each pole is valuable and essential, yet neither can abide alone. Without the complementary opposites, there can be no space for light to emerge!

Like all of life's solutions, it can sound so very easy on paper. The war isn't because of a lack of knowing what to do. The war is overcoming all the physical and circumstantial evidence contrary to our hope that sways our emotions. Having believed something in our hearts, the *war is now in our souls*; it is the struggle between thought and emotion!

We pursue our miracles like children on their way to a dream vacation. The constant question rings out from the backseat: "Are we almost there?" The ability to understand the process and identify where we are in the pursuit of our miracles is vital to our ability to keep hope alive. As long as hope—the confident expectation—remains active, we will remain unwavering (patient) in our journey! Sometimes discouragement is able to rob our fuel—hope—simply because we do not know where we are in the process. We could be one breath away from experiencing our miracles and yet give up, thinking that none of this was working for us!

The Process of Pursuing

In the beginning of the miraculous there is either need or desire. Need or desire is the birthplace of every miracle. Unfortunately, for those who do not know how to live in the miraculous, need or desire can lead to negative emotions like lack or desperation. Some resort to complaining, lying or manipulation to get what they want. But those who embrace the miraculous realize that desire is a miracle-seeking manifestation!

After desire, the next step is persuasion! Persuasion is where we process beliefs, overcome the deception of secondary gain, move past limiting beliefs, confront worthiness issues, gather evidence and persuade our hearts. Faith must have evidence and must overcome all the beliefs (doubt) that make us disqualified! This is where our personal belief system is confronted and renewed. Hope sustains us through the process of persuasion!

If all we had to do was believe for the miracle, it would be easy to experience some of the most incredible phenomenon. The great struggle comes from the very beliefs that brought us to the place of need. Remember our discussion of secondary gain? The reasons we have for keeping our problems may be far more compelling than the potential gains that accompany our miracles. Or, because of personal failures, we may feel completely undeserving of a miracle! Perhaps our sense of self-worth has created boundaries. There is no end to the possible obstacles that must be encountered and overcome to persuade our hearts that the miracle is ours!

If we attempt to solve all our problems before pursuing the miraculous, we will never begin the journey. Plus, mankind is totally incapable of looking inward and evaluating which problems should be solved. Wheat and tares are metaphorical symbols representing that which is wholesome and that which is destructive in our hearts. Wheat and tares look alike until the wheat bears fruit. The plants are very difficult to distinguish. In our attempt to pull the tares out of our hearts, we likely may pull out the wheat. In the end we do as much damage seeking to fix ourselves as what the original problem had done.

Fruit (results) is the only way to distinguish wheat from tares! This can be done only in the process of the journey! For example, if I desire a certain thing and I must walk to town to purchase it, then I have a journey before me. If I sit at home and imagine every possible limitation to my making that journey, I will be discouraged before I start. If, on the other hand, I begin the journey, I will discover the hindrances en route.

Fruit (results) is the only way to distinguish wheat from tares.

When I encounter the barrier, I have the encouragement of the progress I've made to cheer me beyond the current circumstances. At this juncture, I easily can do some heart work to identify why I am stuck at this point. I can release the limiting belief and move joyfully on my way. The fruit of a limiting belief becomes easily identifiable when it manifests itself. Because it is clearly seen as a hindrance to my desires, I will gladly release it!

Simply attempting to write the end you desire on your heart will bring the emergence of limiting beliefs. *Do not be discouraged* when they come! Instead, realize that now you can finally see what has hindered you in this area. And even more wonderful is the fact that you can release them! Negative thinkers see the emergence of limiting beliefs as a sign that they cannot have the miracle they desire. Positive, hopeful thinkers see it as part of the persuasion process.

"Are We There Yet?"

One of the many unanswerable questions I hear asked so often is, "How long will this take?" The amount of time and effort required to persuade your heart is in direct proportion to the reality and emotional persuasion of the current situation. If you are getting a lot of secondary gain from your need, then you have to attach enough pleasure to the desired end to far outweigh what you are now experiencing. If you are terribly afraid of confronting your current situation, you will have to generate enough confidence to wipe away the fear. If there is intense pain in your body, being healed must become more real than the pain! The effort you put into persuading your heart of a new reality must be passionate and consistent enough to outweigh every emotion associated with the problem!

As previously mentioned, you must avoid creating new boundaries by limiting the number of *impossibility* factors you introduce into the mix. The more you add impossibilities to the equation, the harder it is to persuade yourself. For example, if you include another person in your miracle, and that person expresses incongruent behavior, it is hard to stay persuaded. Every unnecessary detail you include limits your ability to believe, thereby increasing the propensity to give up. Only include in your personal miracle that over which you have absolute authority and choice: yourself!

Getting miracles for others requires completely different parameters not suitable for this context. It is not that you should never pursue miracles for others. Rather, you should not include others in your own miracle. To do so complicates your end with theirs. The potential for discouragement is far too great! To achieve success in believing with others, you should be thoroughly acquainted with the differences in application of these laws of the miraculous.

The terms *affirmation* and *confession* are somewhat synonymous. For the sake of clarity I will use the word *confession*. Confession is used more often among a more religious group. But in the end they both are used in the same way: as a declaration of reality. But for the sake of identifying where you are in the journey, let's create workable definitions. First, a confession is the declaration or admission of facts. The ancient Greek word that we translate as confession means to say the same thing. So in the application of the laws of the miraculous, a confession or affirmation is when I am saying the same thing with my mouth as it occurs in my mind, feelings and actions. It is a pronouncement or affirmation of fact.

A persuasion could be the identical statement. However, because my feelings, thoughts and actions are not yet consistently stable, I am not stating a fact of congruence; I am making a statement for the purpose of persuasion. Knowing the difference between a confession and a persuasion can be the difference between despair and hope! If a person makes a statement of persuasion in an attempt to deny reality, the heart will never believe it. It will become a source of internal conflict. Knowing and properly applying a persuasion and a confession (affirmation) is the difference between self-deception and hope!

When we are in meditation or internal dialogue, we always should speak the end before it has occurred. Knowing that we are attempting to align our thoughts, feelings and actions prevents the left side of our brains from accusing us of lying! If we think we are making a confession and there is no internal congruence, then we feel that we are merely pretending! Knowing that we are persuading ourselves makes us feel diligent, persistent and hopeful.

Knowing when we are persuading as opposed to confessing makes us feel diligent, persistent and hopeful.

By observant internal awareness we know when we have shifted from a persuasion to a confession! We feel that very distinct sense of ownership that occurs when hope gives birth to faith. We have a very real awareness that comes from a flood of internal evidence. We peacefully transition from labor to rest. Now when we say the exact words that were previously used to persuade our hearts, there is a feeling of joy, maybe even a slight smile and a new light in our eyes. We have proof of an invisible reality that is completely ours, in the earliest stages of physical manifestation!

At the moment a hope becomes a heart belief, the entire process becomes effortless. The weight of responsibility shifts. The desired end feels like a finished fact. Another evidence of hope becoming a heart belief is that we see and sense ourselves as being the person who is living the new reality. A shift in self-perception is always associated with a new heart belief.

The word *faith*, in the ancient Greek, also was used when speaking of a title deed, that is, proof of ownership. A deed is the proof that you own something that you perhaps have never seen. The deed settles all disputes and quells all emotional competition. When evidence opposing the miraculous end you desire emerges, you simply go back to your deed and all arguments are ended!

Faith, that internal sense of ownership, ends all impatience and struggle. The feelings of joy that normally only appear when physical possession is taken are experienced now because the sense of ownership has replaced the sense of vague possibility. The day you move from hope to faith is the day the struggle ends and rest, peace and joy become the new norm! The more consistently and passionately you persuade your heart, the sooner you can answer the one prevailing question, "Are we there yet?"

Chapter 24

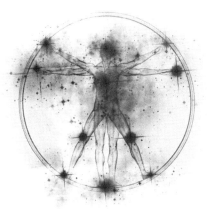

Miraculous Mindsets

According to the Encarta dictionary, *mindset* is defined as "a set of beliefs or a way of thinking that determine somebody's behavior and outlook."[1] One's general outlook is *everything*! It is the filter through which all things are seen, judged, perceived and understood! Our mindset is the precursor to all our actions!

Our outlooks color everything in our world. It paints such a tint on all we see that a pessimistic outlook makes all things seem hard, challenging or threatening, while an optimistic outlook makes all things seem enjoyable, exciting or fulfilling! Albert Einstein was talking about mindset when he said, "There are only two ways to live your life. One is as though nothing is a miracle. The other is as though everything is a miracle."[2] In the end, outlook is often the difference between "I can" or "I can't," the miraculous and the mundane!

Mindset emerges from core heart beliefs. The core beliefs of the heart are the sum total of our sense of self; they are the ultimate guiding

---

[1] Encarta® World English Dictionary [North American Edition] © & (P)2007 Microsoft Corporation. All rights reserved. Developed for Microsoft by Bloomsbury Publishing Plc. <<www.dictionary.msn.com>>, *mindset*. November 28, 2008.
[2] Albert Einstein quotes <<http://www.humboldt1.com/~gralsto/einstein/quotes.html>> November 28, 2008.

beliefs that have integrated into one general mindset that comprises our life paradigms. Everything in our life paradigms is filtered through our sense of self. Therefore, it is of little value to change what we think about anything else, without changing the way we think about ourselves!

It is of little value to change what we think about anything else, without changing the way we think about ourselves!

People with a negative sense of self can change to see the world as full of opportunity. But that would be even more depressing. Now they simply see more opportunity that they believe they can never enjoy. People can begin to see their spouses as wonderful and beautiful. This could hasten a painful divorce. Seeing them as wonderful, they will more quickly push their spouses away. After all, why would they deserve the love of such a wonderful person? Things going "right" hasten the destruction of a person with destructive core beliefs!

The ultimate expression of codependency is when we look outside ourselves in an attempt to meet a need that can be met only in our own hearts. Gregg Braden says, "There can only be one solution to any problem: a change in attitude and in consciousness."[3] The Bible says, "As a man thinks in his heart, so is he."[4] We know that our outer world is programmed from what we believe and think. So we can free ourselves from the codependent temptation to reach out into a world over which we have no control in a vain attempt to solve our problems. Instead, we can

---

3   Motivational Quotes <<http://mqtoday.blogspot.com/search/label/consciousness>> December 9, 2008.
4   See Proverbs 23:7.

go to the one sure place from which inner and outer transformation can occur without force: our hearts!

Our mindsets reflect our habitual state of mind. Therefore, it defines our general, abiding experience of life. A miracle can be obtained to address a particular problem. But without a change of mindset, the problem simply will reoccur. Because it is "repackaged" it may seem like a totally different issue. But upon close observation the roots are found to go back to a particular attitude or belief. Our world and how we experience it is a good indicator of our mindsets. Regardless of how difficult or negative our circumstance may be, the inability to find and experience the good is always a reflection of our mindsets, not our circumstances. A heart that will not align its beliefs with the good things in life has no capacity to find the good! We must believe it is there in order to recognize it when we see it!

Remember, the habitual thoughts of the mind emerge from the heart. Instead of wasting a lifetime fighting with the thoughts of the mind, we must address the beliefs of the heart. As those beliefs are changed, our general mindsets will change. This shift sets off a continuum in the realm of thought, feeling and experience. But the great thing about addressing the core beliefs is, once they are changed at the heart level, the struggle is over. The thoughts, then, that surface from the heart are healthy, positive and life-affirming.

Faith, Hope and Love: Core Attitudes

There are certain core beliefs that influence the hierarchy of life processes. Regardless of your personal beliefs, regardless of being Christian, Jew, Hindu or atheist, the needs are the same! Faith, hope and love comprise the core attitudes that forge miraculous mindsets. The presence, absence or degree to which these are real in your life may manifest in a plethora of individual emotions and actions. But be assured that, regardless of your personal life beliefs, these three—love, hope and faith—are the true masters of your mindset!

Love is, of course, the deepest need of every human being! Some may quickly argue that they would rather be respected than loved. Others would feel they need safety more than love. Some even would desire to be right more than to be loved. And there even would be those who, if they were honest, would want to be accepted more than loved. What each of these and other preferences reflects is not a deeper need than love. Very often these are merely the substitutes we employ to meet the need that only love can meet. Sometimes it is simply the need for these experiences that, once met, meet the prerequisites to feel loved. For example, some would believe they could not be loved if they were not respected. These prerequisites, however, can become destructive, codependent *needs* that prevent a genuine experience of unconditional love!

Love and fear are the two ultimate antithetical states! Every thought, feeling and emotion arises from one of these two opposing core beliefs/states/attitudes. Love is when one feels valued, precious and held in high regard. The enlightened realize that, as children of the Creator, we are loved beyond measure. The deep abiding fear that opposes love is the feeling of rejection and unworthiness, of being nothing or insignificant; all of which culminates in an expectation of punishment or bad things; that is, condemnation. Condemnation then gives rise to doubt, the belief in all opposing negative evidence. This often is seen in pessimism. In its most basic sense, doubt is always rooted in fear and faith is always based on love!

The three opposing core beliefs/states/attitudes are love and fear, hope and condemnation, faith and doubt. Love is the feeling of value and all that makes us feel valuable, safe and self-respected. This gives rise to hope. Hope is very similar to optimism. Hope is the expectation of good things. And hope is the mother of faith. The heart and mind that expect good things, sees good things. It gathers evidence that proves good things. Therefore, it becomes fully persuaded of the possibility of the desired end!

But the three healthy core beliefs always will be challenged by the three antithetical archenemies: fear, condemnation and doubt!

Your mixture of these core beliefs come together to comprise your general mindset. The more deeply rooted you are in love, hope and faith, the more you will embrace and exude the miraculous mindset—the mindset that by default gives rise to the miracles! That's right, there is a mindset that expects and receives miracles as a way of life. For the person with the miraculous mindset, it is practically all effortless.

There is a mindset that expects and receives miracles as a way of life.

People who live in the miraculous tend to express certain attitudes that are virtually impossible to have when core beliefs are skewed. Thankfulness, generosity, mercy, kindness, patience, love, joy, peace, goodness, faithfulness, gentleness, self-control, tolerance and decisiveness are just a few of the lifestyle attitudes and actions that seem to be embodied by those who espouse the miraculous as a way of life.

The casual observer who does not understand the anatomy of the miraculous thinks these people have these attitudes because they are lucky. "Anyone would have those attitudes if everything was going good for them!" they defensively argue. But we know the opposite to be true. These people experience miracles *because* of their mindsets! Apart from a miraculous mindset, rising from healthy core beliefs, a negative person would sabotage every good thing that comes into his life. Such a one's mindset is the antithesis of the miraculous!

The mindset produces the miraculous, seemingly lucky life! Ultimately, the miraculous events and the mindset become a continuum where one nourishes the other, then gives rise to more of the same. This is what Jesus, the Great Miracle Worker, meant when He implied that we will get more of what we have![5] There is no escaping the cycle of our lives unless we have a change of heart resulting in a new mindset. As someone has said, "Faith can move mountains. Doubt can create them."[6] If we are filled with doubt, which is rooted in fear, we continually create new mountains, new obstacles and new limitations. If we are filled with faith, which is the product of love, we always find and experience the very best life has to offer!

Thankfulness and Generosity

Two of the most prevalent mindsets seen in those who live in the habitually miraculous are thankfulness and generosity. The law of sowing and reaping is an emotional law that explains our capacity to give and receive in any area of life. Religious people have confused the law of sowing and reaping. They think, "If I do bad, God does or allows bad things to happen to me. If I do good, God does or allows good things to happen to me." From the outside that is exactly how it appears. But the law of sowing and reaping is not about God doing or not doing. The law of sowing and reaping is about our capacity to give and receive.

There are two planes on which sowing and reaping work. First there is the outer world, the horizontal plane. How we treat people will tend to come back to us in the way others treat us. If we are critical, people will be anxious to criticize us. If we are generous, people will tend to be generous to us.

On the inner plane, however, it is about the capacity to give and receive. If we are not generous, then we will give only when we hope to get something in return. If that is the way we *measure* out our giving, then that is the way we *measure* when others give to us. In this case, if someone

---

[5] See Mark 4:25.
[6] Unknown.

attempts to do something for us, our response is, "What does he really want?" Our only source of measurement is our own motives! If we give to get, then we project our motives onto others.

If, on the other hand, we tend to be generous with no strings or ulterior motives, we have no negative experience when others are generous to us. Generosity is not limited to money. This is about being generous with mercy, kindness, encouragement and love, just to name a few. Those with the miracle mindset of generosity give freely. Therefore, they always can feel the love of others. Those who withhold love or give love conditionally rarely can feel the genuine love that is expressed to them. *Being* loved is of no value when one has little or no capacity to *feel* loved.

When we are generous in any way to those who are not deserving, it becomes easy for us to believe in a generous Creator and a generous universe. Those who struggle with trust may not have chosen the childhood that made trust so very difficult. But they are choosing the adult behavior that makes them untrustworthy. Their untrustworthiness is affecting their ability to believe for a miracle.

The ultimate sowing and reaping occurs in the fact that our view of the world is really our view of ourselves. How we believe and how we *sow* into the lives of others becomes the evidence we have of the world we live in. It even forges our view of God! In all things we see ourselves; it is, therefore, essential that we like what we see! The generous see all things as freely given. They have no difficulty believing in a God or a universe that gives back generously. The laws of the miraculous are the laws they live by. They daily become the miracle in the lives of the people around them.

The person with the ultimate mindset for the miraculous is the person who is truly thankful! Thankfulness is a life paradigm. It is not what you momentarily feel when you get something you want. Thankfulness doesn't fade away. It abides! Truly thankful people look at all of life as a gift. They notice the good things. Because they are thankful, they notice the good things more than the bad things. Thankfulness and optimism go hand in hand! You have to notice the good things to be thankful, and noticing the good things makes you expect them!

The laws of the miraculous are the laws that the generous and thankful live by.

Don't confuse hope, or the expectation of good things, with entitlement! Entitlement feels that good things are owed. If something is owed, then it is a debt and there is little gratitude for a debt paid. The thankful person lives in that rare paradox: "As a child of the Creator, all good things are mine; yet I receive them all as gifts." Realize that an earth that can sustain life, flowers and plants that grow; rain showers, sunrises and sunsets; a body that heals itself; and an infinite number of simple realities are not works of chance. They were part of the loving plan of a benevolent Creator, which makes every moment in life a miracle and a gift!

Grateful people have the desire and joy of expressing gratitude to everyone, all the time! They see every human being as a brother or sister, a child of the Creator to whom they can express their general gratitude for life. If they only have one thing going right in their lives, they are happy for that one thing. That happiness is more the object of their focus than a sky darkened with impending doom.

Generous, thankful people have to obtain the miraculous to sustain the kind of world they believe in. They do not know how to live in a world that is not good—good for themselves and good for others! Their generosity drives them to create a better world and their thankfulness compels them to believe it is possible!

Miraculous Mindsets

The miraculous mindset begins with a change of mind, not a change of circumstances. And like all changes, the first step is choice. Make the choice to be positive, optimistic and loving. To the degree you commit to give these attitudes to the world you create the capacity to receive them. The capacity for a miraculous mindset can begin with your next decision!

Chapter 25

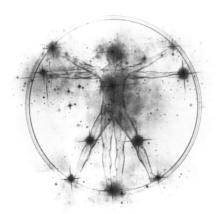

If You Can

Some of the most powerful yet challenging words ever spoken were, "If you can...all things are possible to him that believes!"[1] Millions of people who have hoped to live in the miraculous have pondered those words. They were spoken by the greatest miracle worker ever to have lived. Regardless of our personal beliefs, it would be irresponsible at best to miss the learning opportunity of a lifetime!

Possibly every miracle worker throughout the ages has given us some rendering of this same statement: All things are possible if we believe. Within the pages of this book we have presented the scientific and empirical evidence to support this statement: All things are possible. I have not included the thousands of testimonials from people whom I personally know or who have written to me over the years about how they experienced every imaginable physical healing, operated the law of attraction, overcame death, came back from comas—and an endless list of miraculous phenomenon!

---

[1] See Mark 9:23.

But the phrase I want to consider out of Jesus' statement is, "If you can." There are people who *can*. They "have the heart" for something just as there are those who do not. That doesn't mean those people can't develop the heart for it; it just means at that moment they don't have the heart attitude and beliefs to experience a particular miracle today! Neither does it mean that they cannot believe and receive in the future. This may be the distinction between instantaneous miracles and those that occur through a process…the ability to say, "I can today!"

The Responsibility Is Ours

Jesus made this classic statement when a man brought his epileptic son to Him to be healed. The man began by saying, "If you can do anything, help us." At that moment there seemed to be that all-too-common tendency that we all experience when we need a miracle. He wanted *Jesus* to be responsible for what occurred. The father did not feel that it was within his realm of possibility to do anything. But Jesus immediately placed the responsibility back on him: "No, if *you* can…." If the miracle is for us, then we are the ones who must believe.

Regardless of our religious backgrounds, there is always that tendency for us to want the responsibility for the miraculous to be on the deity of our choice. Using the teachings and model of Jesus, let's examine the level of personal responsibility we play in experiencing the miraculous. We may discover that it is our tendency to want God to do it for us that actually prevents it from occurring!

Jesus said, "Whoever says to this mountain, 'Be removed and be cast into the sea,' and does not doubt in his heart, but believes that what those things he says will be done, he will have whatever he says."[2] Our religious tendency is to read things into this statement that He did not say! Most people would read this statement like this: "If I believe and

---
2   Mark 11:23 NKJV™.

do not doubt, I will speak to the mountain and *God will make it move.*" But that is absolutely not what Jesus taught or modeled. He said, If you believe, speak and do not doubt, it will happen! There was no mention of God having to intervene or take action! This is man functioning as he was designed to function.

Jesus wanted us to know that we are capable of entering into and living in the miraculous at all times. There is no need for us to sit idly by, waiting for the Creator to take action and thereby prove His willingness. His willingness was determined in the way all things in the universe were created and how they operate. This is an interactive world, one in constant interplay with the intelligent life that inhabits this realm. That is proof of His willingness!

The question is not, "If God can." The question is, "If I can." Can I believe that all things are possible and that what I say will come to pass, when I believe it? Do I really believe in the immutable laws of the universe? Do I really believe that I live in an interactive world that is responsive to my thoughts and beliefs? Do I believe that I have authority (the right) to function in this world on the same creative principles as my Creator?

The question is not, "If God can." The question is, "If I can."

Whether you persuade your heart through scientific evidence, empirical observation or religious teaching does not change how the laws work. As far as living in the miraculous goes, how you reach the conclusion is not as important as the fact that you gather your evidence, fully persuade your heart and reach the unshakeable conclusion.

Let's look at another interesting anomaly in the way Jesus brought about miracles. He did not pray long prayers. There was no ceremony. He never asked God to work the miracle. More times than not the conversation would go like this: "I'm crippled; I want to walk!" Jesus would reply, "Then get up and walk!" No formality. No prayer. Simply a command for the person to take action, based on his own decision and his own beliefs.

There are only fourteen individual healings of Jesus recorded in the New Testament. These individual events give us very clear insight into the anatomy of a miracle. Nine of these fourteen times Jesus very specifically pointed out to these people that it was their faith, their deep conviction, their being fully persuaded that produced their miracles. He never laid claim to any special powers beyond those granted to all children of the Creator. And Jesus seemed to always want the people to know it was their faith, not His. He wanted them to understand that the power to work miracles was with them all the time!

Predisposed to Disbelief? Start Persuading!

We live in a society that is very anti-miraculous. That which cannot be "scientifically" explained is rejected. Even what is scientifically explained will be rejected if a single scientist argues against it. Because of the unbelief of previous generations, we were genetically programmed to resist, disbelieve and even discredit the miraculous. Our negativity has become a self-fulfilling prophecy. We've convinced ourselves that the miraculous is not possible. We came to believe it was not possible and, true to the laws of the miraculous, what we believed became our experiential reality! Friend, if you are living in a developed country, you must accept that you are genetically predisposed through generations of pessimism, pseudo-intellectualism and miraculous antagonism to disbelief!

When we see or hear reports of miraculous occurrences in under-developed nations, we write it off to ignorance and superstition! When we hear an intellectual person from a developed nation talk of miracles, we generally discredit him. Personally, I could care less what facilitates a person getting a miracle. If ignorance and superstition play into a person's capacity to involve himself with the miraculous laws of a loving Creator and relieve his suffering, I'm all for it! However, we live in a new day! Ignorance and faith are no longer lumped into one state. Cutting-edge scientific evidence gives support to what was at one time considered ignorant superstitions.

You may have an incredible need for a miracle in your life. If it was put to you at this moment, "You can have this…if you can believe for it!" and your heartfelt answer was an unequivocal, "I cannot believe for my miracle right now," do not deny where you are at this moment! Do not pretend to believe. And above all, do not condemn yourself. Begin persuading your heart. Gather evidence. Faith is the evidence of what you cannot see. Gather enough evidence to overcome all the evidence to the contrary! Gather enough evidence to overwhelm your senses! Gather so much evidence that nothing else makes sense.

It matters little how your miracle comes or how long it takes. It just matters that it comes!

You may not have a spontaneous moment where you are instantly and miraculously delivered from your problem and ushered into your solution. But that's all right! It matters little how it comes or how long it takes. It just matters that your miracle comes! You may not be able to say, "I can believe for it all today." But day by day in increments that grow in sync with your faith, you can say, "I can!"

In 1983 I walked out of a lifelong struggle with a congenital kidney disease. The epic ending to a lifetime of pain and sickness did not consummate in one sweeping experience. Over a four-year period of harrowing struggles in the face of pain, weakness and monumental circumstances, the least of which was a lifetime of knowing, believing and saying, "I have a kidney disease," I persuaded my heart!

Earlier in 1981 I had made a vow to myself. I determined that I would devote a portion of my time every day to investing in my heart. I wanted to be completely, intellectually and emotionally persuaded in the unconditional love of my God and Creator. I determined that I would renew my faith in the miraculous every day! Every day I would spend time seeing myself healed, energetic, debt-free and happy. I can point to no specific day and say, "This is the day my healing became a physical reality." Every day as I persuaded my heart, the reality of health and healing became more real to me than the reality of sickness and pain. Each day, to the degree that health and healing became real, to that same degree sickness and pain became a non-reality!

I was like an escapee seeking to flee my captor when at the last moment the long arm of injustice took hold of the collar of my coat. In a desperate struggle for my life, I mustered all the strength available only to realize I could not escape the ironclad grip that held me bound to my past. Then, in a moment of revelation, I had a heart awakening. My strength would never be sufficient to grant my freedom. Rather than rely on my strength, I would simply "slip out of my coat."

The strong arm of the natural world had no real hold on me. It simply held the outer me, the "me" that I had put on through a lifetime of influence and learning. My coat wasn't the real me. It was just the "me" I wore. Then flexibility and adaptability did for me what strength could not do. I simply relaxed and slipped out of that old coat. I left the natural world holding the old me and I used the immutable miraculous laws of the Creator to create a new coat!

Every day as we meditate and choose our lives, health and circumstances, we weave the fabric of our new coats. Every day that we renew our sense of reality we are saying, "I can." When we daily say "I can," we seldom find ourselves accruing the kind of problems that will challenge our faith. Let's get a miracle for all the small things in our lives and we will never need a miracle for a big thing! But if we do need a "big miracle"… nothing is impossible…"I can!"

Chapter 26

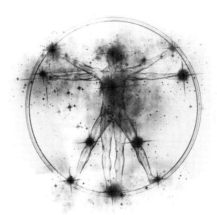

It's Time to Act

To believe is to do! In our Western language we speak of believing as if it were a separate entity, something independent of any corresponding action. In the language of the Great Miracle Worker, *to believe is to act on the belief.* The perfect harmony of belief and action is the final essential key to walking in the miraculous! Nothing can be withheld from the person who thinks, believes, feels, speaks and acts with congruence and consistency—nothing!

You now have the road map marking the way that millions of others have followed to experience their personal miracles. You may have had moments of inspiration during this reading. But all truth is little more than theory until you put it into practice.

At this moment you are compelled to either act or procrastinate. If you act, you have chosen the path of wisdom—the practical application of knowledge. You are on your way to living in the miraculous. If you procrastinate, you need justification. Intellectual justification is one of the most convenient excuses for lack of action! Intellectual justification is unbelief masquerading as the need for more knowledge.

Don't Procrastinate

When proper information is obtained, the only appropriate response is to take corresponding action! All too often we convince ourselves that there is some secret knowledge, some hidden piece to the puzzle or an unknown that keeps us from taking action. The illusion of a mystical unknown is all too often a cloak for our fear and unbelief. The mind all too conveniently creates excuses to defend our fragile self-worth. One of Solomon's proverbs describing just such an inner struggle depicts the person who really doesn't want to take action as surmising, *"There's a lion out there! If I go outside, I might be killed!"*[1]

The pursuit of more knowledge as a substitute for action is common to us all. It is so common that there is an ancient saying about it: "If you think a book can teach you, don't buy the book. If you think a teacher can teach you, don't get a teacher." Once we have knowledge, we don't need more knowledge; we need experiential knowledge. More knowledge hinders and complicates the process if we have not yet actualized what we already have learned! That which is known must be intertwined with who we are. To become real, information needs actualization through application, not more information.

> *Once we have knowledge, we don't need more knowledge; we need experiential knowledge.*

There may be other things to learn, but getting more information without employing what already is known is like piling lumber on a stack of blocks and calling it a house! No! It's just a pile of material that is

---

[1] Proverb 22:13 NLT.

useless until utilized. The first part of a house is the foundation. There is no need to purchase or deliver the wood for framing it until the blocks for its foundation are laid.

The wood lying on the ground waiting for the foundation to be built will rot if it is laid aside for too long! So too is the information that we continuously gather but never use. What is not integrated into our lives by action or meditation within about twelve hours is usually lost. But more than being lost, it actually works against our progress.

When we have learned something that we did not put into practice, it gets categorized mentally as useless information. When the need arises for that piece to the puzzle, we think back and say, "I know that, and it didn't help me!" So instead of utilizing what might be an incredible piece of information, we set off in the pursuit of more information.

Revelation Comes from Action

The revelation for practical wisdom comes with movement—action! We don't know if anything is real until we test it. We can test it through meditation to see how it feels if we pursue a particular course of action. But few people are able to do that type of meditation without interjecting misleading factors! The main way to test anything is to begin putting it into application through small, safe steps! Wisdom is not the depth of insight; wisdom is the practical application of insight. The more we apply, the more we understand!

Revelation is seen as surprising information that is newly disclosed. It's like removing a veil and showing what was previously hidden. But revelation seldom comes to the stagnant. If we cannot see a particular object from our current vantage point, we need not pray for a revelation. We simply should mentally change where we are standing; we should change our opinion or consider another option. The view of the future changes with movement. For example, while standing on the earth, my hand appears larger than the sun. That perception changes the closer I move toward the sun. With enough movement, the opposite of my perceived reality would reveal itself. As we put what we know into practice, we will discover what we need to know. That is a revelation!

As we put what we know into practice, we will discover what we need to know. That is a revelation!

Be comfortable not knowing what to do. Do not surmise that what you know at any point in the journey is all you will ever need! Be comfortable knowing that what you need to know will come to you as long as you keep opening doors and seeking the goal. The New Testament says, "Keep on seeking and you will find; keep on knocking…and [the door] will be opened."[2] People who think they know stop seeking and knocking. Lao-Tzu, considered by some to be the father of Taoism, said, "To know that you do not know is the best. To pretend to know when you do not know is a disease."

If instruction moves you to apply, then you have great faith. If you believe but do not apply, your faith is dead and useless; it has no life. If you do not learn from applying the information that you have, you follow the course of destruction. The wise learn by the application of instruction; the skeptical learn by consequences; the stubborn learn by pain; and the foolish never learn!

You now know what to do. What you don't know is what this will look like in real application. No matter how well it is described, what you

---
2 Matthew 7:7 AMP.

must do in real life cannot be learned from another. No one can teach you what this will look like as you apply it to your life. A teacher can tell how it looked when he applied it in his life. But even when applying the same truth, the subtle variables are vastly different! Ultimately every teacher either leaves his disciple to find the way with the instructions he has given or he foolishly attempts to give what is not his to give: experience!

Here is a review of the instruction you have been given.

- There are no limits, just beliefs. Change a belief and you'll move the boundaries.
- Nothing is too big. The size of the problem is only as big as the emotions you feel about the problem.
- Choose the end you want.
- See it, imagine it and experience it and all its associated positive emotions in every possible way.
- Make the desired end more real than any opposing thought, feeling, circumstance or evidence.
- Never think of the desired end as something in the future; it must be present tense.
- Harmonize your words, thoughts, feelings and actions.
- Eradicate all doubt (belief in an opposing reality) by focusing on the desired end.
- Never fight against doubt and negative feelings. Focus on the desired end.
- Negative feelings will change when the focus is changed.
- Eradicate doubt by reconnecting to the positive feelings associated with the desired end.

- Make your mind work for you, not against you. Change your thoughts and your feelings will change.
- Mental rehearsal molds and grows new circuits in the brain.
- Let you body work for you. The brain's frontal lobe controls other parts of the brain. The image you hold of yourself controls what your brain does to affect the rest of your body. See and speak the desired end.
- Organize your life for the miracle that is on its way.
- Always develop the heart. Beliefs of the heart are at least five thousand times more powerful than thoughts of the brain.
- Daily renew your core beliefs: love, hope and faith.
- Daily release all feelings and thoughts of fear, condemnation and doubt.
- Be thankful about everything. When your miracle manifests, don't feed your ego by taking the credit. Be thankful to a benevolent Creator who created this world and all that exists to make it possible for you to live the life of your choice.

Now it's up to you. Apply today what you have learned…your miracle is on its way! As you apply what you do know, you will come to understand what you do not know. With every step forward you create a new view of the future. Regardless of struggles or opposition, this journey into the miraculous is a wonderful, enthralling, positive adventure! Your life can be a continuous adventure. In fact, you are in danger of smiling more. You may be ambushed by a new, healthy, positive sense of self-worth. You may seem happy for no reason. Life may even become easy and light! You may find yourself adopting my life phrase: Every day's a holiday and every meal's a banquet!

About the Author

JAMES RICHARDS is a pioneer in the field of human development. He has combined spirituality, energy medicine, scientific concepts and human intuition into a philosophical approach that brings about congruence in spirit, soul and body, resulting in a positive, peaceful, productive quality of life. He is a life coach, a consultant, a teacher and a motivational trainer. He holds doctorates in theology, Oriental medicine and human behavior. He was awarded an honorary doctorate for work done in the Philippines. He is a certified substance abuse counselor, a certified detox specialist and a certified herbalist and holds an impressive number of additional certifications and training certificates.

His most noted life's work is Heart Physics®, a life renewal program designed to take people through "Painless, Permanent, Effortless Transformation."

When asked why he has studied such a broad field his answer is simple: "If it helps people, I want to understand it!" The goal of all his work is "to help people experience wholeness, spirit, soul and body!"

Other Books by Dr. James B. Richards:

How to Stop the Pain

Becoming the Person You Want to Be

Breaking the Cycle

Grace: The Power to Change

The Gospel of Peace

Escape from Codependent Christianity

How to Write, Publish and Market Your Own Bestseller

Satan Unmasked

We Still Kiss

Supernatural Ministry

The Prayer Organizer

Effective Small Group Ministry in the New Millennium

My Church, My Family: How to Have a Healthy Relationship with the Church

Taking the Limits Off God

Leadership that Builds People, Volume I: Developing the Heart of a Leader

Leadership that Builds People, Volume II: Developing Leaders Around You

To order products or to contact Dr. Richards, write or call:

Impact Ministries
3516 S. Broad Place
Huntsville, AL 35805
256-536-9402
256-536-4530 – Fax
www.impactministries.com

Positive, Painless, Permanent, Effortless Transformation
essential heart physics

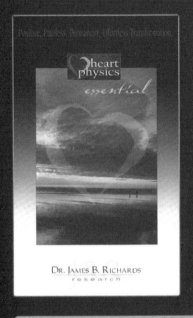

In just 30 days you can discover the ultimate secret to a limitless life! In this life-changing program you will discover the keys to removing your personal boundaries.

Our limitations are nothing more than the boundaries created by faulty beliefs. All we need is the key to open the door that moves us past this lifetime of destructive thinking. All through history the key has represented knowledge, the power of entrance, and unlimited access. In the Essential Heart Physics® Program you will receive the *HeartKey* that, in essence, will give you unlimited access to your heart, bypassing years of faulty logic and reason.

You can simply open the door of your heart and experience:

- A connection to the miraculous.
- Unshakable faith.
- A healthy, positive sense of self-worth!
- The power to change your beliefs at will!
- The unconditional love and acceptance of God!
- Christ, the Great Healer, in you!
- A new confidence!

And…conquer stress, anger, and negative emotions!

Essential Heart Physics® will open the eyes of your heart to see the "unseen" and introduce you to the realm of positive, painless, permanent, effortless transformation.

Order this life-changing program TODAY!
Impact Ministries
256.536.9402 • www.impactministries.com

ALSO AVAILABLE BY MILESTONES INTERNATIONAL PUBLISHERS

TREVOR'S SONG
T.A. Beam

THE MIRACULOUS TRUE STORY OF GOD'S HEALING POWER AFTER A TRAGIC FARM ACCIDENT

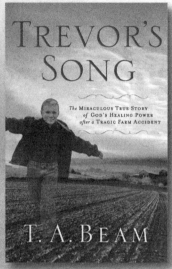

ISBN 978-0-924748-99-8
UPC 88571300069-7

By all odds eight-year-old Trevor beam, horrendously injured in a farm accident, should have lost his right leg and been crippled for life—or so thought the doctors. Yet after only three weeks in the hospital, and less than two months after his accident, Trevor walked into his doctor's office on his own in a walker. Less than two months later he was walking unassisted, completely whole, and without even a limp.

Within these pages you will encounter an amazing story of fatherly love, childlike faith, and the healing power of God. You will meet a little boy who defied everyone's expectations; a little boy who lived when he should have died and who walked again when he should have been crippled. You will meet an earthly father who refused to give up. Most of all, you will meet a heavenly Father who hears and answers prayer; a God who comforted a distraught earthly father with the words, "Trevor is My son too, and I'm going to take good care of him."

• •

BOOKS BY DR. ROGER L. DE HAAN

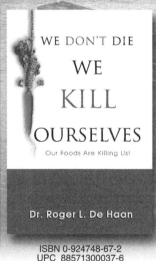

ISBN 0-924748-67-2
UPC 88571300037-6

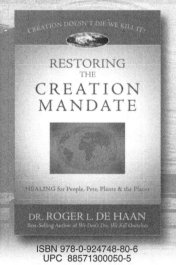

ISBN 978-0-924748-80-6
UPC 88571300050-5

ISBN 978-0-924748-66-0
UPC 88571300036-9

AVAILABLE IN BOOKSTORES EVERYWHERE